Oil Company Divestiture and the Press

Barbara Hobbie

Oil Company Divestiture and the Press

Economic vs. Journalistic Perceptions

foreword by
Richard B. Mancke

Praeger Publishers   New York   London

PRAEGER SPECIAL STUDIES IN U.S. ECONOMIC, SOCIAL, AND POLITICAL ISSUES

Library of Congress Cataloging in Publication Data

Hobbie, Barbara, 1946-
    Oil company divestiture and the press.

    (Praeger special studies in U.S. economic, social,
and political issues)
    Bibliography: p.
    Includes index.
    1. Petroleum industry and trade—United States.
2. Corporate divestiture—United States. 3. Press—
United States. I. Title.
HD9566.H58   1977   338.8   77-10627
ISBN 0-03-022841-7

PRAEGER SPECIAL STUDIES
200 Park Avenue, New York, N.Y., 10017, U.S.A.

Published in the United States of America in 1977
by Praeger Publishers,
A Division of Holt, Rinehart and Winston, CBS, Inc.

789   038   987654321

© 1977 by Praeger Publishers

Printed in the United States of America

FOREWORD
Richard B. Mancke

Any academic who seriously wishes to aid in the design and enforcement of public policy soon recognizes the importance of the press in defining how influential a role he is likely to play. The press's importance derives in large part from the practical (and perhaps psychological) need of elected public officials— those who possess ultimate responsibility for the design of any public policy—to have their accomplishments brought to the attention of the electorate frequently and, on balance, favorably. Since the decline of the electorate's party loyalty and of effective political machines, a politician's success has become even more dependent on the evidence of concern, accomplishment, and credibility that he is able to project. Even a wealthy politician's reelection prospects will be damaged if his public activities fail to attract press attention.

The relationship between politician and press is one of the few truly symbiotic phenomena. The politician needs the press to feed his ego and to confirm and publicize his many talents to the electorate he represents; but the press needs the politician to help legitimize its pronouncements as to what are the true dimensions of the important public issues of the day. One undesirable progeny of this near incestuous coupling of politician and press is the creation of two types of problems for independent experts (that is, those who have relatively detailed knowledge of an issue but lack a vested interest in its outcome) who feel they have a duty to offer their informed advice to aid in defining and then remedying public problems.

The initial problem is one of access. Politicians tend to seek independent advice from those authorities whose opinions can generate favorable publicity. Thus, three sorts of advisers tend to be preferred: the relatively few experts who are already established media celebrities in their own right; those who are experts because they have somehow either been wronged or have witnessed and/or participated in abuses and thus can testify to "facts" that will provoke public attention, sympathy, and/or outrage; and those experts whose advice will confirm their politician-patron's wisdom in soliciting their counsel by supporting— perhaps unwittingly—his preconceptions.

Suppose the independent expert somehow overcomes the first hurdle and attracts political and/or media recognition and attention. A second problem remains—the rather high risk that both politicians and press will either misquote or deliberately distort whatever he has said in order to advance their "higher" ends.

Barbara Hobbie's book uses both content analysis of press articles and statistical analysis of a questionnaire sent to a carefully selected sample of academic economists and newspaper or magazine editors and writers to examine how the nation's periodical press has covered two controversial questions facing U.S. energy policymakers: Is the American oil industry effectively monopolized? Assuming an affirmative answer to the first question, should remedial legislation be passed to force divestiture of large integrated oil companies? Her study concludes that when discussing these questions the periodical press has typically focused on the views of vested interests, either industry spokesmen or industry critics. The press has largely ignored or, worse, distorted the views held by the great majority of academic economists recognized as experts on this subject.

Hobbie's study contains a wealth of detail that should be invaluable to those concerned about the oil company monopoly/divestiture debate. However, it also sheds light on a more fundamental question: Is our highly praised press capable of describing and analyzing objectively important but complicated public policy issues? Generalizing from the results of Hobbie's study, the correct answer to this question seems to be no. In light of the press's key role in defining the scope and content of most current public policy debates, this negative answer has frightening implications.

I disclaim competence to discuss comprehensively the possible explanations for the press's poor performance—as documented by Hobbie—when discussing such a complicated and controversial subject as oil company divestiture. However, one problem that does deserve brief comment is mishandling by the press of the issue of providing balanced coverage of controversial questions. There are at least two classes of problems. First, the media does a poor job of identifying vested interests. Specifically, the press recognizes that the employees or members of formal institutions that have a direct economic stake in a policy's outcome (for example, the involved firm or union) may lack objectivity. But, too often, it fails to recognize that even though the employees or members of public interest groups (for example, environmental or consumer groups, or congressional aides) are untainted by a direct financial interest, they may in fact be no more pristine than those so tainted. One generalization from my experience as a long-time observer of "big oil" is that citizens who have devoted a large portion of their lives to battling either for or against a cause are not especially likely to be objective or fair judges of its merits.

The second problem in promoting balanced coverage revolves around the press's selection of expert sources. The problem was illustrated by J. Frederick Weston in remarks at a recent conference on horizontal divestiture in the oil industry, sponsored by the American Enterprise Institute. (In addition to Weston, the other economists participating were Walter Adams, Morris Adelman, Betty Bock, Darius Gaskins, Richard Mancke, Jesse Markham, Edward Mitchell, Frederick Scherer, and David Teece. With the exception of Walter Adams, all participants, including the present director of the Federal Trade Commission's Bureau of Economics and his immediate predecessor, agreed there was no economic case for horizontal divestiture of oil companies.) According to an unedited transcript of the preceedings, Professor Weston rebutted Professor Adams's earlier reference to the oil industry's economic and political power by stating:

> I submit the best way to test to see who has political power is to look at legislation that gets enacted, and, by that test in recent years, when one looks at all of the punitive legislation against the oil industry, it's clear that that's not where the political power lies. Rather it lies with people like Walter Adams. He is the Kareem Jabbar of government policy. . . . His great oratorical gifts, coupled with irritation from OPEC in the case of the oil industry . . . confers vast powers on Walter Adams. I don't fear the president or chairman of Exxon, in terms of wielding power, I fear Walter Adams and I stand in awe and reverence of him.

No doubt Weston overstates the case. Nevertheless, Hobbie's study attests to its basic truth. Press coverage of oil company monopoly/divestiture issues has tended to neglect the conclusion of most academic experts that the oil companies possess little or no monopoly power. The fact that many of these same economists received significant press attention when they criticized recently abolished federal policies like oil import quotas and the oil depletion allowance (both of these policies strongly advocated by the oil industry) raises the possibility that press coverage of the oil monopoly/divestiture debate has been biased.

Most academics who have studied the oil industry would agree with the proposition that, since at least the early 1960s, the United States has faced several energy policy questions requiring serious and sustained public attention: How can it reduce its still growing dependence on insecure and expensive oil imports? How can it reduce the environmental and health risks attributable to higher production and consumption of oil and natural gas substitutes such as coal, oil shale, and nuclear power? And how can American citizens be guaranteed access to adequate energy supplies without paying unnecessarily high prices? Finding answers to these questions is of vital importance to all Americans. Thus, Congress and the president should have been examining and eventually adopting policies designed to achieve such valuable goals as reducing the monopoly power currently exercised by the OPEC countries, reducing U.S. petroleum demands, and increasing U.S. petroleum supplies. But instead of addressing these fundamental energy policy issues, many of our elected officials have spent the largest portion of the time they devote to energy issues debating what most economists regard to be a nonproblem: how the domestic oil industry should be restructured in order to eliminate monopoly abuses. Given the vital role that the press plays in defining public policy issues, I think that it must shoulder a large part of the responsibility for the costly delays in legislating remedies for real energy problems.

ACKNOWLEDGMENTS

This study was undertaken in conjunction with the Program on International Energy, Resources and the Environment, Fletcher School of Law and Diplomacy, Tufts University. The results of this research also appear in a Master's thesis submitted to the University of Missouri School of Journalism.

The author wishes to thank those economists and journalists who participated in this study and is especially grateful to Richard B. Mancke of the Fletcher School and Keith P. Sanders and Linda Shipley of the University of Missouri Journalism School for their support, critical commentary, and suggestions for revising this book. Thanks are also due to Mrs. Penny Gosdigian for helping to prepare the manuscript.

CONTENTS

LIST OF TABLES

Oil Company Divestiture
and the Press

PAST PERCEPTIONS OF
THE OIL MONOPOLY ISSUE

There has been little love lost between the American public and the oil industry ever since John D. Rockefeller amassed unheard-of wealth by towering over his competitors and eventually capturing more than 75 percent of the oil refining business in the United States. Rockefeller's Standard Oil Trust and public trust in the virtue of big business were gradually broken up through media exposure of questionable business practices, which in turn nurtured growing public suspicion of corporate power, and resulted in a series of corrective antitrust measures. Muckrakers like Ida Tarbell and Henry Demarest Lloyd led the public fight and, in the process, imprinted journalism with the watchdog intaglio.

Monopolies certainly existed prior to Standard Oil of New Jersey, but not until 1911 had any corporation been convicted of violating both substantive sections of the Sherman Antitrust Act, not only by controlling most of the market, but by acting unscrupulously in order to do so. The 1911 conviction, which broke the Standard Oil Trust into 34 units, led to the Clayton Antitrust Act of 1914 and a steady stream of lawsuits and legislative proposals. Nevertheless, charges that the oil industry was monopolized or acted contrary to the public interest continued to be heard over the intervening half century.[1]

The question of what to do about the oil industry was reignited when the Arab oil embargo of 1973-74 dramatized the heavy dependence of the United States on oil imports and the steady post-1970 decline of domestic petroleum reserves. Many Americans were quite willing to believe that the large oil companies were exerting a stranglehold on supply and had invented a shortage as a convenient price-gouging strategy. Others were convinced that these large companies had colluded on any number of oligopolistic ventures, such as squeezing off independent refiners and dealers from supplies, secretly bargaining for lease holdings, withholding lower-cost supplies in favor of insecure and costlier OPEC oil, raising prices arbitrarily to windfall profit levels. These and other similar beliefs, contentions, or suspicions led to public uproar and prompted several senators to move toward a specific form of legislative redress—oil company divestiture.

The first serious attempt to press for oil company divestiture legislation occurred in October 1975. An amendment, attached at the last minute to a natural gas price deregulation bill, would have forced the 16 largest oil companies to divest or separate their vertically integrated operations; it lost by a vote of 54 to 45. A similar amendment to another energy bill lost by a vote of 49 to 40.[2] (The term "vertical integration" refers to single firms engaging in successive stages of the oil business—the production of crude oil, transportation of crude or refined products, refining, and marketing or retailing.) Surprisingly, given the potential impact of such divestiture amendments and the casualness with which they were appended to unrelated bills, the margin of defeat was slim. Several senators, led by then presidential hopefuls Bayh and Jackson, vowed to continue the fight.

Encouraged by the closeness of the 1975 votes, these senators introduced seven bills in early 1976 calling for some form of vertical or horizontal divestiture. (Vertical divestiture would prohibit multiple stage oil operations; horizontal divestiture would prevent oil companies from engaging in other energy businesses such as coal or nuclear.) One bill calling for vertical divestiture of the 18 largest oil companies cleared the Senate Judiciary Committee and would have come up for a vote in late summer of 1976, had key Senate Democrats not been reluctant to have this politically volatile issue introduced into the presidental campaign.

Many view divestiture as a plausible means of reducing what they perceive as the entrenched economic power of the major oil companies and thereby strengthening the U.S. energy position. Yet ironically, the majority of academic economists who are recognized specialists on the oil industry appear to be at odds with popular opinion as expressed by the media, Congress, and various consumer spokespeople. One expert who feels divestiture and similar efforts are misplaced is Professor M. A. Adelman of the Massachusetts Institute of Technology, labeled by many as the leading academic authority on the oil industry, and long a critic of special oil industry privileges. In a *New Republic* review of Christopher Rand's *Making Democracy Safe for Oil*, Adelman wrote:

> Somebody is too powerful but not the companies. The [Arab] governments are clearing over $11 per barrel, even with excessive allowance for costs. To complain about the companies' power is like complaining of a fly in the living room and ignoring an elephant. . . . It also helps [the OPEC cartel] that consumers blame the oil companies and waste time thinking up ways to control or punish them rather than resist the cartel governments.[3]

Contrast this with an interpretation of Rand's thoughts by *Environment* editor Julian McCaull, and the disparity of views about the oil companies' monopolizing abilities (and divestiture-related efforts) becomes clearer and plainer:

> The point made in Rand's book is that the whole basis for pricing of foreign crude oil was historically a contrived one based not on supply and demand but on other sorts of business considerations. . . . Rand details the artificial, inflated basis for that profit margin enjoyed by [oil and gas companies] and now the U.S. public must decide for itself how much it is willing to pay to continue such unusual privileges for a segment of the nation's business community.[4]

Disagreement of this kind and intensity raises some very interesting public policy and communications questions. The inefficacy and disruptive nature of divestiture claimed by its opponents versus the compelling need to dismantle the oil companies' alleged monopoly power called for by divestiture advocates is a critical debate in at least three respects. First, it is important in an immediate public policy sense—if certain economic experts are correct—because if divestiture measures are passed, they could impede the badly needed expansion of domestic energy supplies. Conversely, as those who favor divestiture have suggested, it might lead to positive changes in the way this country develops and utilizes its existing energy resources and envisions future rates of energy growth. That is, a reevaluation of and improvement in U.S. energy policies may emerge as a by-product of the debate over who controls energy development and how much private control benefits the public good. Whatever the outcome of the debate, it has contributed to increased public awareness of energy problems and concern that they begin to be resolved.

The divestiture debate is critical in a second and more subtle way. The immediate issue focuses on major oil companies' alleged control over the oil marketplace (and, to some extent, their abuse of power and privilege in the past, which is believed to have contributed to their current status). But a larger issue is the way in which the public, partly as a result of its attention to the oil industry, will react toward other integrated or large industries. As recent public opinion polls have suggested, anything that connotes big business, corporate power or influence, untaxed wealth, and so on stirs antibusiness sentiments that run quite strong and may even be on the increase. Further, if acceptance of the charge of anticompetitiveness is applied to the oil industry, due to either its integrated structure or its involvement in developing substitute (that is, non-petroleum) energy products, then other industries may be subject to similar charges and potential upheavals.

A final critical factor in the divestiture debate, the one with which this study is most concerned, is whether the press has performed its obligation to the reading public to accurately and fairly depict this important public policy issue. It is especially crucial to examine and evaluate the media's performance on matters of public concern when emotion charges the debate or when factual analyses are complex, as is the case with monopoly/divestiture. A further implication of this study is that the media's performance on this particular issue may also provide some clues as to their abilities to report competently, amidst clamor and seemingly conflicting information, on a broad range of economic issues that directly and vitally affect public policy.

Therefore, this book will first look briefly at the history of muckraking and later criticisms of the oil industry. Then it will examine, in depth, a representative subsection of the media—the periodical press—to evaluate its coverage of the oil company monopoly/divestiture issue. Finally, it will gauge the agreement/disagreement on the issue that exists among economic experts, among different types of journalists, and between experts and the press. Ultimately then, the intent of this study of perspectives on monopoly/divestiture is to clarify the factual basis of the issue, to determine whether the periodical branch of the media has provided the public with a fair and adequate view of the issue, and to examine the broader implications that arise for economic and public policy journalism.

THE CLASSIC OIL MONOPOLY CASE: STANDARD OIL OF NEW JERSEY

Tarbell's View of the Great Anaconda

Few American writers have had the same devasting effects in molding public opinion and forging public action as has Ida Tarbell. The daughter of a school teacher turned tank maker then oil producer, she was immersed from the beginning in the color, raucousness, and fierce competitiveness that characterized the early days of what she called the "oil game." This devout belief in the pleasure of fair battle and the relentless compilation of facts, anecdotes, figures, and folklore enabled her to create a tremendously effective instrument for helping to break up the Standard Oil Trust.

Oily, slithering, and snakelike images abound in her passionate, fact-packed text. Petroleum, first regarded as a nuisance by salt producers tapping saltwater wells, found its initial use as a patent medicine—three tablespoons a day, the first "snake-oil man" recommended. But the potential uses of this readily flammable substance soon excited the minds of adventurous entrepreneurs. The usefulness of petroleum as illuminator and lubricant (only much later as fuel) spurred the cycle of striving to perfect its transportation, refining, and distribution. The first oil was distilled in much the same way as whiskey and, in fact, whiskey barrels and hard drinking rivermen helped get much of the early oil out of Titusville, Pennsylvania and other places in the oil regions via Oil Creek. The wiliest, the most innovative, and the fastest entrepreneurs were those who stood to gain from the rapidly growing industry of oil.

The years following the Civil War were hard times for this infant industry, fraught, Tarbell writes, with various evils. The physical problems of drilling for and transporting oil—for example, building pipelines to replace tank wagons while fending off consequently unemployed and belligerent teamster wagon drivers—required much energy. Various stock swindles bearing the names of war heroes took their toll. The "oil men" (the name Tarbell gave the pioneer producers and refiners, not a derogatory term) had to fight the railroads' practices of freight discrimination while contending with wildly fluctuating prices. All this, Tarbell says, they staunchly regarded as a challenge:

> . . . they looked forward with all the eagerness of the young who have just learned their powers, to years of struggle and development. . . . They would solve all these perplexing problems of over-production, of railroad discrimination, of speculation. . . . They would bring the oil refining to the region where it belonged. . . . There was nothing too good for them, nothing they did not hope and dare. But suddenly, at the very heyday of this confidence, a big hand reached out from nobody knew where, to steal their confidence and throttle their future. The suddenness and blackness of the assault on their business stirred to the bottom of their manhood and their sense of fair play, and the whole region arose in a revolt which is scarcely paralleled in the commercial history of the United States.[5]

The big hand was, of course, John D. Rockefeller, who began as a clerk, graduated to owner of a produce firm, then in 1860 started the small Cleveland refinery that was to swell into the Standard Oil Trust or the Great Anaconda. What Tarbell begins to do at this point in her narrative is to make plausible, through an exhaustive recitation of facts, testimony, and hearsay, the story of how such an assault as Rockefeller's could ever have taken place or been tolerated for so many years.

In essence, Tarbell argues that the Standard Oil Company was guilty of seizing every favorable opportunity that came its way, and if none did, creating opportunities. The oil men failed in their counterattacks against Standard because they did not persevere as diligently, nor stick together when the Trust repeatedly moved to outmaneuver them. This struggle between the independent refiners and Standard Oil began with an insidious spiral. Standard, starting off with an especially good refining process and using highly efficient methods, was able to increase volume enough to persuade the railroads to offer it preferential freight rates. Standard was then able to pick off weak competitors unable to operate under existing freight rates. Thus, Standard took over these companies, increased its volume, and obtained even lower freight rates. As a result it gained greater and greater control of the Cleveland refining market. This gave it sufficient capital at later stages to apply the same sort of procedure to controlling pipelines, crude oil production, prices and supplies, marketing, and refining. In other words, Rockefeller squeezed his competitors gradually but continuously, giving them no room or opportunity for breath between squeezes. He was the epitome of the ruthless businessman.

Tarbell refers to Rockefeller as a "big hand," but he was equally a tight fist. He is said to have controlled his operation to the tiniest detail, even to the point of having the dustbins sifted for usable or sellable materials. His control over employees seems to have been phenomenal as well. He hired many of the presidents of companies he had allegedly run out of business and all Standard Oil employees were reportedly well liked, respected, and listened to by other businessmen in the oil regions. Furthermore, they could be counted upon to exhibit the efficiency and reticence in volunteering information that characterized the man at the top.

But persistence and Spartan zeal were not the Standard's only attributes. Later in its history, Tarbell reports, the company instituted an extensive industrial spy system (often bribing competitors' employees), enabling it to know anything and everything about its competitors' prices, shipping arrangements, costs, and so forth. Thus, Standard was able to engage skillfully in one of its worst offenses—predatory price cutting. Tarbell believed that Standard could drive out competitors with pinpoint accuracy, using both its vast capital and its rigorous information-gathering system. She even tells of a crippled kerosene deliveryman driven out of business by a price-cutting, cart-pushing representative of Standard.

How was Standard able successfully to withstand the repeated attacks by coalitions of other oil men, numerous hostile congressional investigations, court orders rescinding its charters and business arrangements, and scathing, persistent attacks by the regional press? Although Tarbell tries to answer these questions, it is not entirely certain that Standard was invincible.

One explanation for Standard's early success in warding off attacks and allegations is that people tended to forget or ignore individual offenses until people like Tarbell and Lloyd piled them together in large and powerfully written histories, too large for lackadaisical politicians, courts, and citizens to ignore. Another theory, offered more recently and more divorced from the influence of the first muckraking era, is that the Standard Oil Company actually had been affected by these attacks on its practices, that it had already begun to lose its manipulative powers and its hold on market share even before Tarbell published her masterwork.* In all probability, the forces of the press and market adjustments worked together to bring down the Anaconda so emphatically in the Su-

---

*Indeed, Tarbell documents these attacks from all fronts, but tends to dismiss their effects in weakening Standard's position.

preme Court decision of 1911. (The twin-forces theory may be a helpful perspective to remember later when this study addresses the role of the press in changing what may already be a changed situation within the U.S. oil industry.)

Retrospective Views of Standard Oil

History has shown the force and persuasiveness of the early muckrakers' arguments. According to University of Chicago economist George Stigler, "One must regretfully record that in this period Ida Tarbell and Henry Demarest Lloyd did more than the American Economic Association to foster the policy of competition.[6] Nevertheless, later scholarly assessments of Standard Oil suggest that the muckrakers' arguments exaggerated the Trust's monopoly position.

John McGee, at the time a University of Chicago economist, writing in the *Journal of Law and Economics*, states that the purpose of his paper is to determine whether the predissolution Standard Oil Company actually used predatory price cutting to achieve or maintain its monopoly.[7] He writes that settling the question is important to present antitrust policy because the Standard case of 1911 ". . . created a legend. The firm whose history it relates became the archtype of predatory monopoly."[8] McGee goes on to illustrate that although Standard did achieve monopoly in the refining market, it was not by undercutting competitors' prices (and thus losing profits for Standard) but by absorbing other companies through acquisition and merger, measures that were permissible under the law at the time they occurred. McGee contends further that price cutting, where it did occur (most likely at the kerosene peddler level), was highly unsystematic and more importantly, unnecessary. He credits Standard with not indulging in an unnecessary method when a sure one would do.

Another scholar, Henrietta M. Larson of the Harvard Business School, has suggested that Standard's competitors were able, nearly 30 years before the dissolution of the Trust, to recapture segments of the oil industry. "It has been held commonly in the United States that the Supreme Court Decision of 1911 broke the company's monopoly. This is obviously wrong and it greatly oversimplifies and distorts what was a long term development.[9] Larson then discusses a variety of market facts including foreign companies' success in competing with Standard; the strengthening of the Pure Oil Group and the Mellon interests in the East; the competition created by other oil field developments further west, particularly in Texas; the breaking away from Standard by certain marketing operations in the South and midwest; and the losses of economies of scale Standard suffered relative to other companies, once they adopted similar innovations and efficiencies. She also states that Standard was, in fact, sensitive to criticism (especially since it was constantly bombarded with antitrust suits) and that when S.C.T. Dodd became Standard's legal counsel in 1881, his urging of conformity to the spirit as well as the letter of the law was taken to heart. Larson concludes:

> Before the Supreme Court decree of 1911 dissolved the Standard
> Oil combination, competition in the oil industry had established it-
> self on a level of large-scale, integrated operations. No company had
> come to equal the strength of Standard Oil but several were highly
> dynamic and were agressively challenging it in many markets. . . .
> [These competing businesses], the over-all planners and coordi-
> nators, the big risk takers were the larger integrated companies.[10]

Harvard business historian, Arthur M. Johnson, has also expressed the
view that the Standard Trust was weakened by attacks at local and state levels,
although not as substantially as at the federal level. Johnson says that competition, aided by large new sources of petroleum in Texas, began to flourish toward
the end of the nineteenth century and that Rockefeller's competitors began to
mimic the successes of the Great Anaconda.

> With the opening of the Texas fields, a flock of new integrated com-
> panies patterend after Standard Oil soon changed the monopoly
> picture into one of oligopoly. However, the new companies could
> not escape the inheritance that earlier industry efforts to use public
> policy as a tactical weapon had bequeathed to them. . . .
>   In turning to public policy, oilmen had been grasping a
> double-edged sword. While their intention was for it to cut only one
> way—at Standard Oil—it could just as well be used against them.[11]

Thus, the Interstate Commerce Act of 1887, the Hepburn (pipeline) Act
of 1906, the Sherman Antitrust Act of 1890 and other federal legislation began
to be invoked against companies believed to be conducting business unfairly or
squelching competition within an industry. The case of the Standard Oil Trust,
according to Johnson, became something of a scapegoat for all the abuses the
public suspected large corporations of perpetrating.

> . . . Theodore Roosevelt embodied perfectly the ambivalence of
> popular attitudes toward big business. He never ceased to stress
> the importance of big business to the economy, but he distinguished
> between "good" and "bad" combinations on a subjective basis. He
> found in Standard Oil an ideal whipping boy to achieve his legislative
> and political ends. . . . His primary emphasis was on Standard's be-
> havior, which he compared with a system of morality that would
> justify everything from ballot-box stuffing to murder. At the same
> time, he condemned the literal language of the Sherman Act as run-
> ning counter to the economic forces of the age and asked for execu-
> tive and administrative discretion in dealing with big business.[12]

In sum, the clear, hard-edged villany of the Standard Oil Trust becomes
somewhat fuzzy over time, blending into a rather schizoid public image of large

oil companies as both necessary and untrustworthy. The designation of monopoly power is not as simple as it was in the early days of oil, when it involved questions of honor, street-brawling manhood, and hearty, free-spirited competition. Tarbell felt that the Standard Oil Trust put a stop to all that. Her view that Rockefeller may have had his greatest effect not in his specific deeds, but in molding the tenor of big business—frugal, efficient, tight-lipped, relentless, uncompromising, and insatiable—may be quite accurate.

No doubt, current attitudes toward corporations in general have been greatly influenced by this seminal event in business history. But, as the next section suggests, other developments—notably, the interwining of big government and big oil—have undoubtedly contributed to current animosities and to the call for divestiture or other measures to curb suspected oil industry excesses.

GOVERNMENT'S RELATIONSHIP TO THE OIL INDUSTRY

The government's responses to the oil industry have, since the early 1900s, illustrated a sort of "right hand doesn't know what left hand is doing" philosophy. For example, while the Federal Trade Commission was attacking oil company mergers, acquisitions, alleged trade and price-fixing agreements, and other monopolistic or anticompetitive measures, and while the Federal Power Commission was holding natural gas prices artificially below the cost of alternate fuels, other segments of government were formulating policies that advanced the interests of the oil industry. Three basic policy devices were especially important: state market demand prorationing, oil import quotas, and special tax advantages. The effect of the first two policies was to hold oil prices substantially above world price levels during the 1950s and 1960s. The tax policies, especially the oil depletion allowance, allegedly led to higher after-tax profit rates for the oil industry than for other industries.

State Market Demand Prorationing

Originally, market demand prorationing was adopted by various oil-producing states in order to prevent overdrilling of fields with closely spaced wells; it was both a conservation and market stabilizing measure. Because petroleum lies under pressure in underground pockets or pools, it migrates to the wells of those who release this pressure through drilling. The "rule of capture" recognized the right of any person owning land over an oil pocket to produce as much oil as he could, often draining neighboring producers' oil from the same underground sources, reducing pressure in the reservoir, and ultimately leading to lower total production. By 1916 the rule of capture was recognized as causing both economic and physical waste.[13]

In 1924, President Coolidge became sufficiently concerned about wasteful overdrilling on both public and private lands to create a Federal Oil Conservation Board to

> . . . study the government's responsibilities and to enlist the full cooperation of representatives of the oil industry. . . . The oil industry itself might be permitted to determine its own future. That future might be left to the simple workings of the law of supply and demand but for the patent fact that the oil industry's welfare is so intimately linked with the industrial prosperity and safety of the whole people. The Government and business can well join forces to work out this problem of practical conservation.[14]

The production problem worsened considerably during the Board's lifetime from 1924 to 1932, especially after the development of new oil supplies in the enormous east Texas field coincided with precipitous, Depression-related drops in demand. This prompted the Board to recommend voluntary cutbacks in production. When these proved inadequate, the oil states legislated mandatory prorationing laws that restricted the maximum rate of output allowed each well. A section of the National Recovery Administration Act, known as the Connally Hot Oil Act, further assisted state prorationing efforts by prohibiting interstate shipments of illegal, nonprorated or "hot" oil or refined products.

In 1969 all of the academic economists called to testify before the Senate Subcommittee on Antitrust and Monopoly agreed that the ultimate effect of prorationing, from its inception through the 1960s, had been to reduce domestic oil production and thereby to elevate the price of domestic oil far above competitive levels.[15] Prorationing also failed to do what it was created to do—prevent costly overdrilling in already badly depleted fields. Those fields that had deeper, more closely spaced wells were awarded higher allowables or production quotas. And low productivity stripper wells, which produced less than ten barrels per day, were exempted completely from prorationing.

Oil Import Quotas

The primary purpose of the 1969 Senate Subcommittee on Antitrust and Monopoly hearings was to critically evaluate what was then regarded as the domestic oil industry's most important protection device—oil import quotas.

In providing a background for the committee, the former editor of the *American Economic Review*, Paul Homan, argued that three factors led to oil import quotas.[16] The first was a post-World War II deficiency in U.S. oil supplies, made more severe by earlier inefficient or suppressed oil production. Second, after the war, there was also an increase in consumer demand for oil. The third factor was a steep price rise. These led to

. . . a burst of exploratory and development activity which, outrun-
ning demand, produced a large excess of producing capacity. . . .

At the same time, rapid development of oil resources else-
where in the world, especially in Venezuela and the Middle East,
began to outrun demand in foreign markets. In this situation, the
international oil companies began to look to the American markets,
and imports into the United States began to rise rapidly, favored by
the lower cost of foreign oil. The domestic industry greatly resented
this trend and mounted an active campaign for restrictions.[17]

The government, too, feared that increasing oil imports would impair the country's national security by making it too reliant on foreign oil. When voluntary
restrictions by the international companies failed, President Eisenhower established a mandatory oil import program in 1959.

Eventually, however, it became clear that restrictions on oil imports pre-
vented competition from much cheaper (until 1973) foreign oil. The high oil
prices provoked the dismay of consumers in regions that produced little or none
of their own oil. In 1969 President Nixon established the U.S. Cabinet Task
Force on Oil Import Controls to study the efficacy of the oil quotas. This task
force concluded that oil import controls cost American consumers about $4.8
billion in 1969 alone.[18] The report criticized the quota program on the basis
that

[it] is not adequately responsive to present and future security considerations. The fixed quota limitations that have been in effect for
the past ten years, and the system of implementation that has grown
up around them, bears no reasonable relation to current requirements for protection either of the national economy or of essential
oil consumption. . . . The present system has spawned a host of
special arrangements and exceptions for purposes essentially unrelated to the national security, has imposed high costs and inefficien-
cies on consumers and the economy, and has led to undue government intervention in the market and consequent competitive distor-
tions. . . . A majority of the Task Force finds that the present import
control system, as it has developed in practice, is no longer acceptable. The basic question, then, concerns the character and degree of
import restriction judged necessary to safeguard the nation against
severe economic weakening or supply deprivations.[19]

Tax Advantages

The Oil Depletion Allowance

Originally, legislation providing oil depletion allowances was passed to
encourage oil exploration. A crude-oil-producing firm was allowed to substract a

percentage of its total crude oil production from its taxable income. The premise was that companies would want to produce more oil in order to obtain more allowances, but in practice the measure was far from successful. Furthermore, as some members of the House of Representatives argued in a proposal to lower the depletion rate:

> In recent years the petroleum industry, which reports on corporate tax returns, has paid taxes to the United States and foreign governments amounting to 21 percent of taxable net income. This compares with an average tax of over 43 percent paid by manufacturing companies not eligible for various kinds of special tax treatment.[20]

Thus, the depletion allowance percentage was revised downward from 27.5 to 22 percent by the Tax Reform Act of 1969. And in 1975, the depletion allowance was abolished completely for all major producers.

Other Tax Benefits

At least two other tax policies benefiting the oil industry provoked public concern. One, intangible drilling expensing, gave oil and gas producers the option to deducting as current expenses nonequipment outlays for wages, repairs, fuel, hauling, and other intangibles associated with drilling operations. This enabled producers, in effect, to defer making tax payments on revenues from successful drilling ventures (and to deduct the costs of drilling dry or unsuccessful holes.[21] Thus, although producers had to pay taxes eventually, they could reinvest or earn interest on their deferred taxes for several years.

Another practice that has been questioned is the foreign tax credit. This enables companies producing oil overseas to receive a tax credit, applicable to their U.S. corporate income tax, that is actually based on tax payments they have made to foreign oil-producing or host nations in lieu of royalties. (A royalty is a per-barrel charge companies pay in exchange for the right to produce a foreign country's oil.) Understandably, there is disagreement as to whether the foreign tax credit constitutes a subsidy to multinational oil companies or shields them from a form of double taxation.

SUMMARY

Three basic policies—market demand prorationing, oil import quotas, and the oil depletion allowance—aided the oil industry either by allowing American producers to charge above-world-market prices or by giving them a tax subsidy. However, due to world market conditions and domestic decline in production, the benefits these policies provided had begun to diminish in the early 1970s, and by 1976 all three were defunct. First, because U.S. oil output had been dropping since 1970, prorationing no longer significantly restricted oil production. Second, oil import quotas were abolished by a presidential executive order in 1973. And third, the oil depletion allowance was abolished for all except small oil companies by the Tax Reform Act of 1975.

In spite of the elimination of these significant three privileges, the oil industry appeared to the general public to be more prosperous than ever. Following the 1973 embargo, oil prices soared. Independent refiners and retailers complained that they were squeezed or denied access to cheaper domestic oil that only the major producers could supply. Renewed cries of monopoly began to fall on eager ears. It was at this time, as the next chapter relates, that the move for divestiture of the major oil companies' integrated operations became full-blown.

NOTES

1. A 1976 *Newsweek* article contends that public distrust of big business extends to other than the oil industry ("Embattled Businessman," *Newsweek*, February 16, 1976, pp. 56-60). However, public distrust of oil companies seems stronger than public distrust of other large industries such as automobiles and steel. A series of *Wall Street Journal* articles, under the general title "Big Oil Besieged," illustrates the depth of present anti-oil emotions (*Wall Street Journal*, February 9, 11, and 12, 1976).

2. For elaboration see Arlen Large, "Congressional Outlook for the Breakup Plan: Wait Till Next Year," *Wall Street Journal*, February 11, 1976, p. 1.

3. M. A. Adelman, review of *Making Democracy Safe for Oil*, by Christopher T. Rand, *The New Republic*, November 15, 1975, p. 26.

4. Julian McCaull, "Letters: The Author Replies," *Environment*, November 1975, pp. 42-43.

5. Ida M. Tarbell, *The History of the Standard Oil Company: Briefer Version*, ed. David M. Chalmers (New York: Harper and Row, 1966), pp. 20-21.

6. George Stigler, "Monopoly and Oligopoly by Merger," *American Economic Review* 50 (May 1960): 30-31.

7. John S. McGee, "Predatory Price Cutting: The Standard (N.J.) Case," *Journal of Law and Economics* 74 (Fall 1958): 137-69.

8. Ibid., p. 137.

9. Henrietta M. Larson, "The Rise of Big Business in the Oil Industry," in *Oil's First Century*, ed. the staff of *Business History Review* (Cambridge, Mass.: Harvard University Press, 1960), p. 38.

10. Ibid., p. 42.

11. Arthur M. Johnson, "Public Policy and Concentration in the Petroleum Industry, 1890-1911," in *Oil's First Century*, ed. the staff of *Business History Review* (Cambridge, Mass.: Harvard University Press, 1960), pp. 51-52.

12. Ibid., p. 55.

13. M. A. Adelman, *The World Petroleum Market* (Baltimore: Johns Hopkins University Press, 1972), p. 44.

14. U. S. Senate, Committee on the Judiciary, Subcommittee on Antitrust and Monopoly, testimony by Robert Engler (quoting President Coolidge), *Government Intervention in the Market Mechanism: on S. Res. 40*, 91st Cong., 1st sess., 1969, p. 365.

15. Ibid.

16. Ibid., p. 104.

17. Ibid.

18. U.S. Cabinet Task Force on Oil Import Controls, *The Oil Import Question* (Washington, D.C.: Government Printing Office, 1970), pp. 259-63.

19. Ibid., p. 128.

20. U.S. Congress, House, *H.R. Report Number 91-413*, 91st Cong., 1st sess., 1969.

21. Richard B. Mancke, *The Failure of U.S. Energy Policy* (New York: Columbia University Press, 1974), p. 79.

2

BLAME-LAYING RENEWED:
A CLASH OF PERCEPTIONS

Following Senator Henry Jackson's March 1976 presidental primary victory in Massachusetts, an editorial in the *Wall Street Journal* lamented that:

> The emergence of Scoop Jackson as a viable candidate poses a painful dilemma for those of us who believe he is profoundly right on questions of foreign policy and profoundly wrong on questions of economics.
>   . . . in this agonizing there is an especially sour element . . . his use of the oil companies as whipping boys for the oil crisis caused by the Arabs. . . .
>   Now a certain exaggeration, flexibility and even demagogy is normal in politics. . . . But precisely because of his record and stature, one expected better of Scoop Jackson. It is precisely because Scoop knew better that the episode looms so large.[1]

Although Jackson did not survive the rigors of the presidential primary campaign, his attack on the oil companies drew a burgeoning following in public opinion polls and among growing numbers of his congressional colleagues. The point made by the *Journal*, that the oil companies have been used as whipping boys, would probably be met with even greater skepticism today than in the preembargo days of the early 1970s when the oil companies enjoyed numerous special governmental privileges but considerably less public exposure. It was then that the severest energy woes of the United States began.

Beginning in late 1972, a year before the OAPEC embargo, the United States was hit by shortages of domestically refined petroleum products. As a result, refined product prices began rising rapidly, and retail marketers of unbranded gasoline or fuel oil and independent refining companies complained of being cut off from supplies. Independents in several states brought suits against major oil companies. These complaints prompted the Federal Trade Commission to start investigating the business practices of the large oil companies.

Senator Phillip Hart, chairman of the Senate Subcommittee on Antitrust and Monopoly, had suggested in 1971 that the FTC begin investigating the majors. However, it was not until June of 1973 that the FTC's actions were publicly aired and even hurried along a bit. In a conspicuously issued letter to the FTC chairman, Lewis Engman, Senator Jackson asked the FTC to make a 30-day crash study of charges that the oil companies conspired to create the refined petroleum product shortage. Jackson wrote:

> The very fact that the so-called "conspiracy theory" is supported by circumstantial evidence and that it does have credibility among knowledgeable observers of the industry makes an in-depth investigation mandatory. . . . If investigation shows that the current shortages have been calculated and engineered for private profit or advantage, legislation will be required to effectuate fundamental changes in the structure and operation of the petroleum industry.[2]

By July of 1973 there was considerable talk among members of Congress and within the Federal Trade Commission of reorganizing the petroleum industry's structure. Testifying before the Senate Subcommittee on Antitrust and Monopoly, James T. Halverson, director of the FTC's Bureau of Competition, stated that his staff "has preliminarily concluded that the structure, conduct, and performance of the petroleum industry raises serious antitrust issues."[3] The report by Halverson's staff (which, contrary to FTC practice, had been leaked to reporters prior to his testimony) was said to recommend divestiture or the separation of the major oil companies' crude-oil-producing operations from refining, transportation, and marketing. Later that month, the FTC issued a formal complaint charging that the eight largest domestic oil companies—Exxon, Texaco, Gulf, Mobil, Standard of California, Standard of Indiana, Shell, and Atlantic Richfield—had monopolized refining.[4]

After the start of the eye-opening OAPEC embargo in October 1973, the debate about the extent of oil companies' abuses and anticompetitive practices grew more vigorous and dichotomized. A number of senators (Kennedy, Hart, Moss, Abourezk, and others) aligned themselves with the FTC's Halverson in charging that the oil companies had, among other things, withheld natural gas supplies to raise wellhead prices, underreported oil and gas reserves, squeezed independents by denying them access to crude oil supplies through price fixing and preferential treatment to large refiners, controlled pipelines and therefore crude supplies, held down refinery capacity, and worked continuously to maintain favorable government policies such as oil import quotas and the oil depletion allowance.

Other segments of government disagreed with these varying charges and assessments of the extent to which oil monopoly reigned. For example, the Federal Power Commission defended the petroleum industry's competitiveness in sales of natural gas and argued that reporting of reserves by industry had not been deceptive. Also, a Treasury Department staff report strongly supported by the president's oil policy committee chairman, William E. Simon, attacked the Federal Trade Commission's report. Simon claimed it was inaccurate, unfairly biased against the eight named oil companies, and detrimental to an expansion of domestic refinery capacity. An exchange of argumentative letters between Simon and FTC Assistant Administrator Robert J. Lewis, drew front-page attention from the Washington *Post*, which suggested that the White House had intervened inappropriately in the FTC proceedings.[5]

There was disagreement as to the degree of major oil company monopolization. A spokesman for the Antitrust Division of the Justice Department opposed a total divestiture proposal offered in late 1973 by Senator Frank Moss, arguing that "[W]e believe there may be sound reasons for enacting legislation which would require that oil pipelines be independently owned. . . . But apart from the special case of pipelines, we cannot presently support an absolute prohibition of vertical integration in the petroleum industry."[6]

It must have been extremely difficult for many of the undecided senators investigating the oil industry to sort out different testimonies and determine which sources were most reliable. The confusion was undoubtedly compounded for the general public listening to and reading about so many conflicting views presented throughout 1973-74. While Federal Energy Office administrators were testifying that the oil crisis was genuine, union leaders were denouncing it as a hoax. While independent marketers and refiners filed suits against the majors in some states, the leaders of the independent producers and pipeline owners claimed that the oil industry was being scapegoated and vastly misrepresented.[7] Experts by the dozens were pulled from each detractor's or defender's favorite woodwork as more and more Senate committees entered the fray: the Permanent Investigations Subcommittee, the Subcommittee on Antitrust and Monopoly, the Interior Committee, the Commerce Committee, the Subcommittee on Multinational Corporations, the Energy Subcommittee of the Joint Economic Committee, and the Special Subcommittee on Integrated Oil Operations, to name the most prominent. At one point there was even a committee to decide whether one central committee should amass oil industry information and testimony—a procedural matter no one was ever able to resolve.

By early 1975 the monopoly charge had become a familiar refrain and the issue (perhaps largely because of its complexity) became more politicized, often seeming to diverge along liberal-conservative lines. Stories of unethical or illegal practices by the major oil companies, such as secret deals with producing companies and large, unreported campaign contributions, began growing both in number and magnitude.

Congressional ire at perceived oil company privileges and abuses resulted in several more investigations and in the abolition of the percentage depletion for all but the small, independent producers. As discussed in Chapter 1, prorationing had become ineffectual as domestic output fell, and quotas had been abolished by executive order in 1973. Thus, the percentage depletion allowance was considered by many government leaders and industry critics to have been the last solid bastion of privilege for the major oil companies.

But such a judgment proved premature. Some senators felt that the fight had just begun. A thick atmosphere of distrust of big business enveloped the nation. The OAPEC embargo had shaken U.S. confidence and underlined the severity of energy scarcity. It was also the year before the presidental election; to some candidates charges against the powerful oil companies seemed well worth repeating. And although those senators most adamant in their support of divestiture did not survive the presidental battle, the issue still simmers on the congressional back burner, with the idea of horizontal divestiture gaining the prominence vertical divestiture proposals recently enjoyed.

ECONOMIC EXPERTS' PERCEPTIONS

What would be the effects of divestiture, particularly vertical divestiture?* Would it combat monopoly and revive competition? Or, as some economic experts have suggested, is the monopoly issue a false one and the role of firm integration a red herring? An editorial in *The New Republic*, usually a strongly liberal periodical, summed up some of the confusion and rancor surrounding divestiture in a "might as well get the oil companies" fashion:

> Proposals to break the oil companies into smaller pieces have been around for a long time, but they are more likely to succeed in 1976 than before because people are angry about prices. Although petroleum is cheaper in the US than in any other industrial country, a price jump in one year from 30 cents to 52 cents for a gallon of gas is possible only through monopoly control of the market, and consumers know it. Since it isn't possible to strike at OPEC, the headquarters of price rigging, Americans are ready to attack the nearest surrogate—OPEC's collaborators and partners in this country, the major oil companies. . . .
> Because no other agency will tackle the problem of concentrated power in the oil industry, Congress is left with the job. It's

---

*This study addresses vertical divestiture in particular. However, many of the arguments for or against vertical integration could be applied to a discussion of horizontal integration.

not the kind of job that Congress does well, for legislators and their staffs don't have the time or expert knowledge to perform the delicate surgery that's needed to bring competitive life back into the energy business. But it's the only surgeon in town.[8]

If Congress had either time or expert knowledge, would it reach the preceding conclusions? To arrive at such conclusions, economists who specialize in the study of industrial organization and practices are apt to look at different and more confined sorts of evidence than are legislators who must deal with public concerns and answer to their constituents. Economic experts, in Senate subcommittee testimony and in scholarly works, have considered several types of evidence. How this evidence has been filtered down and translated to the general public forms the major subject of concern in this study.

ECONOMIC EVIDENCE OF MONOPOLY[9]

Economists often examine three types of evidence to determine the extent to which firms behave monopolistically: market structure of a firm or firms—the ability of competing firms to enter an industry and conduct business competitively, and the degree of market share or control; firm conduct—the extent to which firms in a market behave, either explicitly or implicitly, in collusive or anticompetitive ways (for example, agreeing to fix prices, divide markets, or apply legal or marketing tactics that might harass current or potential competitors); and profit rate—the extent to which firms in an industry enjoy persistently high profit rates (that cannot be explained except by monopoly-type arguments). In addition, in the case of oil companies, experts have examined special arguments, such as the implicit effects of vertical or horizontal integration on market control and competition; the ability of integrated companies to squeeze independent, unintegrated companies as a result of the majors' control over crude oil production or other stages of the oil business; and the effects of special devices like joint ventures and overlapping directorships.

Because the Senate antitrust subcommittee claims to have based much of its decision making on economic (rather than sociological, psychological, or other types of) evidence, it is to the economic experts that we look for an answer as to whether oil company monopoly exists and, if so, whether divestiture is a feasible solution.

Market Share

Market share is one common measure of monopoly or lack of interfirm competition. Of the oil industry, Professor Neil H. Jacoby of UCLA's Graduate School of Management has written:

. . . the U.S. oil industry is *relatively less concentrated* than U.S. manufacturing industry as a whole. Concentration is normally measured by the combined market shares of the top four firms in an industry. The average for U.S. manufacturing (weighted by value added) was 40.1 percent in 1970. In petroleum, the top four firms combined had 27 percent of crude oil production, 35 percent of crude oil reserves, 34 percent of gasoline refining capacity and 30 percent of gasoline sales. Moreover, no one firm towers over the industry. EXXON, the largest, accounted for only 8.5 percent of crude production, owned 11.6 percent of crude reserves and held 9.2 percent of U.S. refining capacity. It was surpassed by Texaco, which has 8 percent of the national market.[10]

Professor Richard B. Mancke, Fletcher School of Law and Diplomacy, Tufts University, argues similarly:

Compared with other natural resource based heavy industries—such as steel, aluminum, or copper—or large manufacturing industries—such as automobiles, computers, or electrical equipment—American crude oil production is not highly concentrated in the hands of just a few giant firms. Instead, there are more than 20 large companies (annual petroleum sales greater than $1 billion) that are significant participants in the crude oil industry. . . . Besides the presence of many large firms, there are literally tens of thousands of small companies that presently compete in the American crude oil industry. The role played by small companies is especially crucial in the vital exploration phase. In 1974 small companies (defined as not in the largest 30) drilled 86.2 percent of all exploratory wells.[11]

There is some disagreement among economists as to what these concentration or market share ratios really mean. Dissenting from the previous two views, economists Joel B. Dirlam of the University of Rhode Island and Walter Adams of Michigan State University stated before the Energy Subcommittee of the Joint Economic Committee:

At first blush, the concentration ratios in crude oil production do not appear to be overwhelming. . . . Even so, it is noteworthy that concentration has been steadily increasing since the mid-1950's so that by 1973 the 8 largest companies accounted for almost as big a share of crude oil production as did the 20 largest in 1955. This trend is largely explained by the massive mergers during this period—especially mergers between the very largest companies.[12]

Adams and Dirlam quote Dr. Walter Measday (majority staff member on the antitrust subcommittee and long an advocate of divestiture) who points out that

. . . concentration in reserve ownership is even more important, particularly for the future, than concentration in current production. And the largest companies control most of the proved reserves. The Federal Trade Commission staff found that in 1970 our sixteen major companies controlled 77 percent of the net proved oil reserves in the United States and Canada. The producer has effective control, however, over all the oil he lifts including the shares for royalty owners and other nonworking interest holders. In terms of gross reserves, the sixteen majors may control more than 90 percent of existing proved reserves.[13]

Economic experts also look at a measure related to market share—the "ease of entry" measure. Industrial organization economists Adelman and McKie, whose quotations follow, express the view that entry into different stages of the oil industry by many firms is relatively easy, lending little credence to arguments that the oil industry bars competing entrepreneurs.

First, Professor James McKie of the University of Texas describes the way ways firms enter one stage, the crude oil business:

Many oil producing companies originated as successful wildcat enterprises. While a few firms may begin with a large supply of capital and immediately undertake an extensive drilling program, the typical firm got its start through a series of fortunate single ventures, often involving exploratory deals with established major or independent firms. New corporations and partnerships are frequently budded from the existing ones. . . .

A geologist or petroleum engineer may gain enough experience on his own, making good use of associations he has built up in the industry. . . . An employee of a drilling contractor may work up from platform hand to superintendent. Once known to purchasers of drilling services and sellers of equipment, he finds it relatively easy to set up his own firm. . . . After operating as a contract criller for some time, he may be willing to put one of his rigs into a wildcat venture on a speculative basis. . . . In this way drilling contractors frequently become independent producers. . . .

Another way to enter oil and gas exploration is via brokerage. Exploration enterprise swarms with middlemen anxious to arrange producing deals. . . . A speculative broker may arrange a prospecting deal among other parties . . . and usually he retains for himself a small interest in the venture. Since technical training and apprenticeship are not strictly necessary, this route is crowded with hopeful shoestring promoters along with the experienced entrepreneurs.[14]

Second, Professor M. A. Adelman of M.I.T. writes of the world tanker market:

Each individual ship available for spot charter is, in effect, like a separate firm and the worldwide market allows no protected enclaves. . . . In any given month, several dozen ships are offered for oil company use all over the world by several hundred owners, none with over 5 percent of total tonnage. Tacit collusion would be impossible, and no attempt at open collusion has been made since World War II. . . . [The] "spot" charter market therefore seems purely competitive.

The time-charter market is linked to the spot market at one end, and at the other to the cost of creating new capacity. Here entry is open and cheap. . . . Moreover, there are no strong economies of scale in ship operations. Many owners have only one ship. .. . But to say that many competent firms cluster on the boundaries of the industry, and that minimum capital requirements are low, is to say that entry is easy and market control impossible.

With many ships available in the short-run, and easy entry for the long-run, what possibility is left for control in the meantime? Little if any in theory, and none can be observed in practice. Tankship owners, oil companies and independents cannot control the long-term supply even in concert, for anyone contemplating a production or refining investment and needing the transport services has time to charter a ship or buy a new one.[15]

Firm Conduct

The ways in which firms choose to do business and interact—firm conduct—is another measure of the competitiveness of an industry. As Mancke has written:

When an industry has many firms and entry is easy, collusive behavior becomes nearly impossible unless all of the industry's major firms are tied together by explicit price-fixing and market-sharing agreements. Such agreements would violate Section 1 of the Sherman Act. There is no evidence that they exist in the American crude oil industry.[16]

Adelman, in a 1976 talk to a Harvard seminar entitled "The Economics of Divestiture," expressed similar doubts that the majors had colluded to fix prices or divide markets.[17] He cited two types of firm conduct that could affect competitive behavior—overlapping directorships and joint ventures. He concluded that collusion by oil company executives at board meetings (other than their own companies') would be rather senseless. Why, he asked, collude only at board meetings? He also stated that although exchange of information about companies' offshore land lease bids probably would be very rewarding to the oil companies, none had been alleged. Further, joint ventures among oil companies bidding for oil lands are not confined uniformly to major companies, nor do the same large companies always dominate bidding. Any scheme of market division, he concluded, would be very complex, and thus, difficult to keep secret.

Adams illustrates a contrasting view of oil company behavior when he states that

> . . . the petroleum majors are intertwined with one another through a seamless web of interlocking control. They do not function as independent or competitive, but as cooperative entities at every strategic point of the industry's integrated structure. They are meshed with one another in a symbiotic relationship which almost inevitably precludes any genuinely competitive behavior.
>
> Joint ventures are one manifestation of this symbiotic relationship. A joint venture established a community of interest among the parents and a mechanism for avoiding competition between them. It provides the opportunity for foreclosing non-partners from access to supplies and/or access to markets.[18]

This view is not supported, however, by a specific study of joint ventures conducted by Professors Edward Erickson and Robert Spann (of North Carolina State University and Virginia Polytechnic Institute and State University, respectively). They find the U.S. petroleum industry to be one of the least concentrated of U.S. industries and add that

> . . . the firms in the petroleum industry engage in a number of joint activities, including joint ventures to bid for and develop offshore OCS [Outer Continental Shelf] leases. We have examined the record of the sealed bid auction market for offshore OCS lease sales from 1954 through mid-1973. We compared the patterns of bidding behavior and the composition of joint ventures with those which can reasonably be expected to have prevailed were the practice of forming joint ventures for offshore lease sales an example of collusive or anti-competitive behavior. The implications of such behavior are that joint ventures would be substituted for solo bids, that joint ventures would lead to stable market shares of OCS tracts won, that the incidence of joint ventures would be positively correlated with firm size, that majors and non-majors would not enter joint ventures together, and that identical bids might be observed.
>
> The facts are uniformly inconsistent with these implications. As the risks and uncertainties associated with offshore exploration became more widely understood, an increased incidence of joint ventures was associated with an increase in the total number of bidders, an increased number of bidders per tract, and a decrease in the relative numbers of tracts which receive only one bid. For smaller firms . . . joint ventures decrease the chance of gambler's ruin in offshore exploration.

Majors have been much more likely to enter joint ventures which contained no other majors rather than joint ventures which contain two or more majors. . . .

Joint ventures, including joint ventures among major and non-major firms, have facilitated entry into offshore activity and have increased the number of bidders. In terms of their effect upon competitive results, joint ventures in offshore OCS lease auctions are pro-competitive.[19]

Firm conduct measures have also come to include arguments about the inherent effects of vertical and horizontal integration in providing some firms with special advantages. The current Federal Trade Commission suit against the eight largest majors, popularly known as the "Exxon Case," is a landmark suit. Instead of charging the companies with specific, traditional antitrust violations, as cited in the Sherman Act, it focuses on the cooperative behavior and integrated structure of these companies. Actual collusive behavior may be very difficult for the FTC to prove. Thus, the suit has paid particular attention to showing that each company's stages (exploration and production, refining and transportation, distribution and marketing, corporate identification and computer operations, and international markets) provide the companies with a network of internally and mutually supportive relationships whose effects are to reduce competition from others and among themselves.

Adelman addressed the FTC divestiture effort directly in his talk on the economics of divestiture. He stated that vertical integration simply does not raise any monopoly problems; that the existence of viable, unintegrated firms, showed that integrated firms were not really necessary. Nevertheless, the transactions costs of divesting would be enormous; in short, "divestiture would be a vast exercise in shuffling papers."[20]

Jacoby, whose views on concentration and market share were cited earlier, agrees with Adelman in one sense. Although he places more importance on vertical integration as a factor in firm efficiency, he agrees that competition from unintegrated firms has been strong:

. . . there are thriving *un*integrated firms at every stage of the petroleum industry, and they account for a substantial share of the market. The *un*integrated firm has its own strengths, such as a management which specializes in single-stage operations and is particularly knowledgeable about local and regional markets. In fact, unintegrated refiners and marketers have, as a group, expanded their market share at the expense of the integrated companies during the past twenty years.

The U.S. oil industry *has relatively less vertical integration* than has U.S. manufacturing as a whole. . . .[21]

Professor Edward J. Mitchell of the University of Michigan, in testimony before the Senate antitrust and monopoly subcommittee, amassed information on a number of economists' views on the incipient effects of vertical integration. According to Mitchell:

> [A] look at the economic literature on vertical integration shows that there is no presumption that vertical integration weakens competition. The majority of writers believe that vertical integration has no weakening effect on competition and indeed that it reduces the harmful effects of monopoly where monopoly does exist. These writers include Bork, Liebler, Peltzman, Spengler, Schmalensee, and Warren-Boulton. A smaller number, notably Edwards and Mueller, believe there are some circumstances where vertical integration permits anti-competitive actions, such as market "squeezes." But even these writers do not find vertical integration presumptively anti-competitive. As Corwin Edwards puts it, vertical integration "tells nothing either about power or abuse of power. Hence it implies neither monopoly nor absence of monopoly. In so far as the monopoly problem is concerned, it is a neutral term."[22]

The FTC complaint offers the related argument that vertical integration facilitates squeezing of independents. The complaint argues that a noncompetitive market situation has been exacerbated because the majors "[pursue] a common course of action accommodating the needs and goals of each other in the production, supply and transportation of crude oil to the exclusion or detriment of independent refiners and potential entrants into refining . . . [and] marketing."[23]

A similar argument can also be found in the works of University of Indiana marketing professors Fred C. Allvine and James M. Patterson who served as advisers to the Jackson Permanent Investigations Subcommittee. They argue that the vertically integrated majors retain dominance by using monopoly profits from crude oil production to finance practices that eliminate independent marketers. Such practices by the majors include running refining and marketing operations at a loss, thus undercutting independents' prices via price wars; artificially increasing prices at which they transfer their own crude to their refinery divisions; and cutting or squeezing off refiners (and therefore, downstream marketers, jobbers, and station dealers) from supplies of crude oil or refined products; or setting prices that, in effect, cut off supplies.[24]

Even in 1972, when oil import quotas and the depletion allowance were still in effect, there was disagreement as to the validity of Allvine and Patterson's line of argument. C. David Anderson wrote in a 1973 *Yale Law Journal* review:

> While the Allvine and Patterson position has a certain appeal, on closer inspection a number of theoretical flaws appear. For example,

they offer no explanation of why a company which has already captured the available monopoly profits at the crude oil level would benefit by using these profits to obtain a monopoly at the refining and distribution levels. Nor do they explain why competition between the majors themselves—and there are at least twenty of them—has not dissipated the alleged excessive profits in crude production.

Their argument is more directly exposed by concentrating on their evidence that a crude monopoly profit exists and is used as a downstream subsidy. First, consider the structure of the industry. One can agree with Allvine and Patterson that federal import restrictions and state prorationing have combined to maintain domestic crude oil prices above the competitive level. But although such regulation has been generally effective in preventing competitive pressure from lowering the prices for crude, there is no system to prevent competition from raising the price of oil properties. Subsidies like the depletion allowance should therefore be expected to disappear into higher bonus payments, state taxes, and the costs of drilling marginal prospects.[25]

Anderson further points out that most majors buy considerable amounts of crude oil from both major and independent competitors. Their control over crude is, he notes, only about half of all crude oil production, rendering the majors' incapable of squeezing independents as Allvine and Patterson argue.[26]

Mancke, too, in an extensive paper for the American Enterprise Institute's National Energy Project supports Anderson's view that the depletion allowance (import quotas were no longer in effect) in no way enabled the majors to control crude oil pricing, shift profits to noncrude operations, and thus engage in squeezing.

The FTC analysis was wrong because it failed to take proper account of the fact that most of the integrated majors were not self-sufficient in crude oil. To operate their U.S. refineries at desired levels they had to buy crude oil from independent producers. Assuming that the oil depletion allowance was 22 percent, profit-shifting would only yield profits for those companies able to produce at least 93 percent of their crude oil needs. Except for Getty Oil, only the sixteenth largest, none of these integrated giants produced more than 93 percent of its total domestic needs. . . . The after-tax losses [of profit-shifting] . . . would have ranged from a low of 3 cents on each dollar of profits shifted by relatively oil-rich Marathon to a high of 48.3 cents on each dollar of profits shifted by relatively oil-poor Standard Oil of Ohio. None of these 16 integrated majors would choose to bear these high costs. This implies that, even if it were possible, profit-shifting would never be practiced and thus that independent refiners would never be "squeezed."[27]

Profit Rates

Another popular argument, one that probably would be recognized as the most gut-level argument, is that the major oil companies enjoyed a sudden burst of windfall profits during the early months of the OAPEC embargo and continued to report high quarterly earnings well into 1974. The dispute that erupted among various experts and government and industry officials concerned whether these earnings were justified or exorbitant and whether they were temporary or persistent, that is, what reported earnings actually meant for long-term oil industry profitability trends.

Economists Erickson and Spann, examining a common measure—the after-tax rate of return on equity investments—conclude that most U.S. oil companies' profits were below the average for all U.S. industrial firms between 1963 and 1973.[28] They suggest that the sharp post-embargo rise in oil company profits was due primarily to much higher crude oil prices coupled with temporary inventory and accounting-related profits. It was these idiosyncratic, temporary windfalls that produced the most public stir and confusion. Suspicions of profits hidden or disguised by accounting department mechinations provided the media with a number of irate congressional quotes and conclusions.

Mitchell examined an alternative profitability measure, the return actually received by oil company common stockholders between 1953 and 1974.* On the basis of these data he concludes:

> An investment of $1,000 in the American international oil companies or 14 domestic refiners or 2 domestic producers whose stock was listed on the major stock exchanges would have left the investor worse off than an investment of $1,000 in the S & P [Standard and Poor's] 500 Stock Composite Index. Oil company stockholders have not reaped monopoly profits. Indeed, they have fared somewhat worse than the average owner of common stock.[29]

This chapter has tried to present a number of arguments that have been used either to endorse or reject the notion that major oil companies have achieved a monopoly position within the oil industry and/or that divestiture of the integrated operations of these companies would help to restore competition. Although the chapter deliberately presented arguments that attest to either view, it

---

*Mitchell subtracted the sum of the stock's purchase price at the start of the period and all dividends paid during the period from the value of the stock at the end of the period. In his calculation he assumed, for purposes of assessing the highest investment return possible, that all dividends that had been paid to oil company stockholders were reinvested in the same common stock.

should be noted that the majority of the economic experts examined have failed to find the alleged anticompetitive practices. Thus, the statement made by Professor Mitchell, in testimony before the antitrust subcommittee, seems to sum up many of the major arguments that economists who have studied oil industry structure and behavior seem to be making. His fairly extensive discussion warrants quotation also because of its seeming disagreement with many of the views, or at any rate doubts, that have been publicly and widely expressed by non-economic experts.

> I appear today to discuss the wisdom and consequences of forcing vertically integrated petroleum companies to divest themselves of billions of dollars of assets and of reorganizing the U.S. petroleum industry into separate producing, refining, transportation, and marketing companies, each operating at only one stage in the industrial chain.
>
> The ostensible purpose of this coerced reorganization is to make the petroleum industry more competitive and, thereby, lower the price consumers pay for oil products. For this divestiture to have the desired consequences three conditions must be met:
>> (1) the industry must be monopolistic or less than competitive as it stands, with articifically high prices for oil products;
>> (2) the monopolistic elements in the industry must be contingent upon vertical integration (otherwise vertical divestiture would not destroy the monopoly); and
>> (3) the benefits of increased competition induced by divestiture must more than offset the loss of the economies of vertical integration.
>
> For vertical divestiture to be sound public policy *each* of these three premises must be valid. If any one is invalid the argument for divestiture fails.
>
> If he is to be taken seriously, an advocate of divestiture must develop strong arguments based on economic logic and objective facts that each of these premises is valid. I believe it is impossible based on the current state of knowledge to establish *any* of the three premises with respectable economic research.
>
> A number of questions or issues bear directly on the divestiture issue and the validity of the three premises. I have dealt with many of them in my research and my conclusions are as follows:
>> (1) Vertical integration is not presumptively anti-competitive.
>> (2) Senator Bayh's statement that "no other industry is so completely vertically integrated" is incorrect; the petroleum industry has a relatively low degree of vertical integration.
>> (3) The petroleum industry is competitive.
>> (4) Vertical divestiture would not make the petroleum industry more competitive.

(5) There is no evidence of monopoly profits in the petroleum industry.

(6) There are substantial cost savings from vertical integration including lower operating costs and lower capital costs.

(7) Small non-integrated firms are not squeezed out of the industry and entry by new firms is not difficult.

(8) Vertical integration is not confined to large privately-owned American oil companies; it is common among both small and large, government-owned and privately-owned companies throughout the world.[30]

ARE THE EXPERTS AT ODDS WITH THE MEDIA?

In looking at the statements made by a number of respected authorities on oil industry structure and performance, many puzzling discrepancies and questions arise. Why, if the economic experts have credibility, is there so much anti-industry sentiment and why has this focused so intensely on the oil industry? Were these experts paid more attention in the late 1960s when they objected to special oil industry privileges, such as prorationing and import quotas, than they are now? Why have traditional measures of competition been scrapped by government litigators and legislators? (Because the oil industry is different? Because this is the only way to prosecute successfully?) Why did price and profit skyrocket so suddenly? Who is to be believed?

Perhaps these are some of the factors that have influenced the media in their presentation of the issue. Certainly, a pivotal problem this book hopes to resolve, or at least illuminate, is whether the press is able to identify experts who can provide answers to complex economic questions or whether the press is actually interested in seeking out experts whose opinions may clash with the media's perception of what is happening.

The intent of this study, then, is to do some conceivably painful probing into whether that portion of the media being studied has performed with integrity in presenting the oil monopoly/divestiture issue. While it is reasonable to expect public charges, suspicion, or possible intrigue to outweigh the dryness of academic exposition—especially in times when many so-called facts have shown themselves to be wide open to intrepretation, or even to be fabrications or illusions—it is also reasonable to expect the media to present as accurate and complete a picture of a situation as can be derived from the available factual base.

Chapters 4 through 6 will attempt to examine precisely what it is the press has said about the oil industry, its structure, its competitiveness—based on economic or other criteria—and the tone and tenor of the media's current thought about the industry, from big oil to independent. In order to examine the various facets of this issue, these chapters will employ a formal content analysis method, described in Chapter 3. The subjects for this analysis are those news and opinion periodicals that have covered the oil divestiture/monopoly issue in some detail since the start of the OAPEC embargo. The content analysis will concern itself particularly with those sources (both people and scholarly or statistical sources) and arguments that have been cited or incorporated into these media presentations. The use of sources and arguments should provide numerous clues as to the extent to which these periodicals—in choosing which facts to present—have channeled readers' opinion-forming processes. The content analysis will also examine any slant or advocacy directly presented in individual articles or in a periodical's presentations as a whole.

NOTES

1. "Scoop Emerges," *Wall Street Journal*, March 4, 1976, p. 12.

2. "Probe of Energy-plot Charge Sought," *Oil and Gas Journal*, June 11, 1973, p. 68.

3. "FTC Staff Would Divest Majors of Production," *Oil and Gas Journal*, July 2, 1973, p. 13.

4. "FTC Moves to Break Up the Big Eight," *Oil and Gas Journal*, July 23, 1973, p. 29.

5. "Treasury Flails FTC Divestiture Action," *Oil and Gas Journal*, September 10, 1973, p. 57.

6. "Justice Supports Oil-line Divorcement Only," *Oil and Gas Journal*, December 24, 1973, p. 14.

7. "Charges of U.S. Refining Monopoly Hit," *Oil and Gas Journal*, March 18, 1974, p. 46.

8. "Breaking Up Oil," *The New Republic*, December 27, 1975, p. 8.

9. Some of the material in this section results from the author's work as a research assistant to Richard B. Mancke (Associate Professor of Economics and director of the energy studies/research program at the Fletcher School of Law and Diplomacy, Tufts University) in preparing a background paper for the Twentieth Century Fund's Task Force Report on U.S. Energy Policy, *Providing for Energy* (New York: McGraw-Hill, 1977), pp. 93-108.

10. Neil H. Jacoby, "Vertical Dismemberment of Large Oil Companies—A Disastrous Solution to a Non-Problem," prepublication article accepted by the New York *Times* based on Jacoby's testimony before the Subcommittee on Antitrust and Monopoly, Committee on the Judiciary, U.S. Senate, February 18, 1976, pp. 2-3.

11. Richard B. Mancke, statement before the Subcommittee on Antitrust and Monopoly, Committee on the Judiciary, U.S. Senate, January 22, 1976 (xerox copy of testimony prior to publication), pp. 5-6.

12. Walter Adams and Joel B. Dirlam, statement before the Energy Subcommittee, Joint Economic Committee, U.S. Congress, December 8, 1975 (xerox copy of testimony prior to publication), p. 2.

13. Ibid., pp. 2-3.

14. James McKie, "Market Structure and Uncertainty in Oil and Gas Exploration," *Quarterly Journal of Economics* 74 (November 1960): 569.

15. Adelman, *The World Petroleum Market*, op. cit., pp. 105-6.

16. Mancke, Subcommittee on Antitrust and Monopoly testimony, op. cit., p. 8.

17. M. A. Adelman, talk before Harvard University seminar, Economics 2590, April 5, 1976.

18. Adams and Dirlam, op. cit., p. 3.

19. Edward Erickson and Robert Spann, "Entry, Risk Sharing and Competition in Offshore Petroleum Exploration," (unpublished manuscript, North Carolina State University, December 1975), p. 33-34.

20. Adelman, Harvard seminar, op. cit.

21. Jacoby, op. cit., p. 2.

22. Edward J. Mitchell, statement before the Subcommittee on Antitrust and Monopoly, Committee on the Judiciary, U.S. Senate, January 22, 1976 (xerox copy of testimony prior to publication), p. 3.

23. *Federal Trade Commission Complaint*, Docket No. 8934, pp. 8-9.

24. Fred C. Allvine and James M. Patterson, *Competition Ltd.: The Marketing of Gasoline* and *Highway Robbery: An Analysis of the Gasoline Crisis* (Bloomington, Ind.: University of Indiana Press, 1972 and 1974, respectively).

25. C. David Anderson, review of *Competition Ltd.: The Marketing of Gasoline*, by Fred C. Allvine and James M. Patterson, *Yale Law Journal* 82 (May 1973): 1359.

26. Ibid., p. 1362.

27. Richard B. Mancke, "Competition in the Oil Industry," in *Vertical Integration in the Oil Industry*, ed. Edward J. Mitchell (Washington: American Enterprise Institute for Public Policy Research, 1976), pp. 65-67.

28. Edward Erickson and Robert Spann, "The U.S. Petroleum Industry," in *The Energy Question*, ed. Edward Erickson and Leonard Waverman, 2 vols. (Toronto: The University of Toronto Press, 1974), 2: 6-12.

29. Mitchell, Subcommittee on Antitrust and Monopoly testimony, op. cit., p. 60.

30. Ibid., pp. 1-2.

CHAPTER

3

DIVESTITURE, MONOPOLY, AND
THE PERIODICAL PRESS:
CONTENT ANALYSIS METHODOLOGY

WHAT'S SOURCE FOR THE GOOSE

The sleeper ascendency of Jimmy Carter onto the Democratic presidental ticket offered a temporary suspension of the divestiture debate. Senator Bayh reluctantly deferred a floor flight on the politically prickly issue prior to the fall 1976 recess of the 94th Congress. President Carter has already influenced the shape and direction of current divestiture efforts, for he appears to prefer horizontal to vertical divestiture (with strict antitrust enforcement for vertically integrated oil companies). However, what a predominantly Democratic Congress will ultimately decide about divestiture seems to be more in the hands of the political fates than a matter of subcommittee groundwork.

An illustration of the slipperyness of the subject happened at the Democratic National Convention in New York. NBC floor reporters interviewed a Texas delegate and an Energy Action spokesman, actor Paul Newman. And each, respectively, asserted that, as president, Jimmy Carter would oppose/favor breaking up the major oil companies' integrated operations. In this instance, the presentation of sources with opposing views was balanced (although Newman's high recognition factor may have given him a competitive edge on a memorability scale). What the event accentuated is that source selection (one variable of communication content) is an important component in determining and delimiting the information newspeople are able to present.

Thus, this study will examine in depth not only sources, but a number of other variables in order to evaluate the coverage given monopoly/divestiture by one branch of the media—the periodical press. The method to be employed is that of content analysis. The next section describes some of the risks and ad-

32

vantages of such a procedure and the following section discusses how the method has been applied to this particular study.

THE USE AND MISUSE OF CONTENT ANALYSIS

Numerous researchers have extended and refined the definition of what content analysis should be and do. In a classic work, social scientist Bernard Berelson termed content analysis "a research technique for the objective, systematic, and quantitative description of the manifest content of communication."[1] Fred N. Kerlinger in *Foundations of Behavioral Research* defined content analysis as "a method of studying and analyzing communications in a systematic, objective and quantitative manner for the purpose of measuring variables."[2] He adds,

> Most content analysis has not been done to measure variables, as such. Rather, it has been used to determine the relative emphasis or frequency of various communication phenomena: propaganda, trends, styles, changes in content, readability . . . content analysis, while certainly a method of analysis, is more than that. It is . . . a method of observation.[3]

Phillip J. Stone and his associates Dunphy, Smith, and Ogilvie, define content analysis as "any research technique for making inferences by systematically and objectively identifying specified characteristics within text."[4] This definition diverges from Berelson's in that it considers the inference-making process to be a far more central function of content analysis. As these authors note:

> This is not a trivial change from past definitions for it imposes upon the researcher the burden of integrating theory with method, of being dissatisfied with mere description of phenomena. This does not restrict the researcher to strictly hypothesis-testing studies, but it does require him to construct categories that he believes are relevant, either singly or in combination, to his conceptual framework.[5]

Despite this major difference, these authors and others who have made a study of content analysis agree on a number of basic ground rules.

Basic Requirements in Using Content Analysis

A very basic requirement is that the method fit the problem. Because it consumes much time and effort and because it poses certain problems in the areas of category construction, recording, reliability, validity, interpretation, and so on, content analysis is often recommended as a last-resort technique. Nevertheless, it has been recognized as adaptable and useful in many instances. It can be used when large quantities of textual materials applicable to a subject are readily available. And it has been found useful in studies whose objectives are to compare content with other forms of communication content, to compare content with certain standards of performance or values, to evaluate changes in content over time, to discover certain characteristics of a body of content, to reveal focuses of attention, to identify concerns and possibly the intent of the communication, or to perform other systematic inquiries into what has actually been communicated and in what way.

Lasswell, Lerner, and Pool regard content analysis as a means of going beyond ordinary research.

> Content analysis should begin where traditional modes of research end. The man who wishes to use content analysis for a study of the propaganda of some political party, for example, should steep himself in the propaganda. Before he begins to count, he should read it to detect characteristic mechanisms and devices. He should study the vocabulary and format. He should know the party organization and personnel. From this knowledge he should organize hypotheses and predictions. At this point, in a conventional study, he would start writing. At this point in a content analysis he is, instead, ready to set up his categories, to pretest them, and then to start counting.[6]

Another requirement of content analysis is that it be systematic, objective, and quantitative. This oft-repeated cardinal rule clearly defines the primary purpose of content analysis: to accurately characterize content in a way no other methodological approach can. The raison d'etre of content analysis is that it can record and evaluate what is said, the way something is said—and often, the implicit message of a verbal statement, more precisely than a casual perusal, a randomly selected reading, a survey, a scale, or any other method of examination. Performed properly, content analysis is also an objective and impartial measure of content. This distinguishes this methodology from other forms of analysis that admit conjecture, inference, or personal opinion on a freer basis.

Another facet of content analysis is its quantifying function. Content analysts count a variety of variables and units of measurement (for example, words, whole articles) in order to derive information about the nature, direction, intensity, emphasis, and so on of a given body of content. Counting, in exemplary content analysis, is purposeful and fruitful. However, counting must be developed within a sound theoretical framework. As Cartwright warns,

> One of the most serious criticisms that can be made of much of the research employing content analysis is that the "findings" have no clear significance for either theory or practice. In reviewing the work

in this field, one is struck by the number of studies which have apparently been guided by a sheer fascination with counting.[7]

The proper application of the specifics of content analysis to a research problem is the most taxing requirement but the one that has been least specified. This is because the analysis is so dependent upon the problem or hypothesis at hand, the design of analytical elements, and the degree of specificity a researcher desires. It is in this sense that the method is a subjective one. It involves first defining a universe of possible choices and then choosing among those many options. From the onset of a study, the researcher's knowledge and judgment are critical factors affecting the outcome of a content analysis.

The Elements of Content Analysis

One of the first things a content analyst must decide is which content elements to quantify. As Stone et al. write,

> The content analysis procedure involves the interaction of two processes: The specification of the content characteristics to be measured and the application of rules for identifying and recording the characteristics when they occur in the data. The categories into which content is coded vary widely from one investigation to another and are dependent upon the investigator's theory and the nature of his data.[8]

The content characteristics, units, or elements (whichever one chooses to call them) can include one or more of the following: words, themes, characters, items, space-and-time measures, images, plot composition, and patterns of beliefs and arguments.* For most practical purposes, unless the researcher is able to conduct a computerized scan and counting of textual material, the word is a difficult unit to employ. Aside from this objection, the word still might be useful to identify emphases, biases, prejudices, linguistic peculiarities, or other such characteristics.

A number of content analysts have agreed that the theme is one of the most useful elements. However, it is sometimes necessary to break down and reconstruct a theme because it can appear in varied and nonuniform sentences, assertions, statements, phrases, slogans, propositions, arguments, and so on. Berelson regards the theme as especially useful in studies of the effect of communica-

---

*The first five are Berleson's; the latter three—images, plot, and patterns of belief and arguments—are categories suggested by Stone et al.

tion upon public opinion because "it takes the form in which issues and attitudes are usually discussed."[9] Stone et al. have taken Berelson's concept of theme a step further:

> Berelson defines a theme as "an assertion about a subject matter." We roughly divide such assertions into two varieties. One employs an active verb and a noun to describe the event, the noun being either a subject or an object of the verb. Examples include "maltreat minorities," "soldiers revolt," "dictator attacks," "underpay workers," as well as complete transitive statements, such as "shopowners underpay workers." A second form is a static assertion described by a word or its modifiers, such as "nasty neighbor" or "healthful exercise.". . .
>
> Within this book, we use the *theme* primarily to refer to an assertion involving events described by active verbs. The more static assertions, implicit in many noun-modifier relationships, are called *images*.[10]

But in most manual or noncomputer analyses, the researcher might be willing to sacrifice some accuracy by relaxing both Berelson's and Stone's limitations on the actual form theme must take, especially since it is regarded by both as useful in studying political situations and attitudes in content.

Characters or plot elements are also used in content analysis studies dealing with literature, drama, and related subjects.

The element most frequently used in content analysis, according to Berelson, is the item, "the whole 'natural' unit employed by producers of symbol material." He goes on to say, "The item differs among the different media: it may be a book, a magazine article or story, a speech, a radio program, a letter, a news story, an editorial or any other self-contained expression. . . ."[11]

Another element is the space-time measure. This could include measures of the number of column inches, lines, or space devoted to a topic, or time measures of nonprint media coverage.

Patterns of beliefs and arguments also find their way into many content analysis studies. In the 1930s these elements were often limited to the famous propaganda devices—glittering generalities, name calling, bandwagon, card stacking, and so on. Subsequent studies have looked at a much wider range of beliefs and arguments in exploring human logic systems, cognitive abilities, cross-cultural differences, and other varied phenomena.* This is a difficult area to pin

---

*This category has been used, for example, in studies of cognition by R. Abelson in 1963 and of neurotic logic by K. Colby, also in 1963. Both studies first reduced arguments to simplified sentences for purposes of computer analysis.

down, Berelson says, because it "usually involves a whole complex of content, and is thus difficult to analyze objectively."[12] However, it would seem that this objection could be overcome if the subelements of patterns of beliefs and arguments could be properly broken out for study.*

What element or elements should be selected for a particular study? Berelson answers this with several suggestions: more than one unit of analysis can be profitably employed in a single study: word counts, items, and space-time measures are most appropriate for studies of straight subject matter emphasis; large units ordinarily provide as valid an analysis of direction (approval-disapproval) as small units; themes take more time to analyze than simple symbols or large units for the same reliability; and the theme is an appropriate unit for studies of meaning and of relationships among meanings.[13]

Selecting Categories in Content Analysis

The selection of variables in content analysis affects the validity of a study (whether or not it measures what it purports to measure), the reliability of results (whether or not results are reliable, stable and dependable, and whether or not similar studies can reach parallel conclusions), and the accuracy of hypotheses and predictions—in short, the study's credibility as a whole. As Berelson puts it,

> Content analysis stands or falls by its categories. Particular studies have been productive to the extent that the categories were clearly formulated and well adapted to the problem and to the content. Content analysis studies done on a hit-or-miss basis, without clearly formulated problems for investigation and with vaguely drawn or poorly articulated categories, are almost certain to be of indifferent or low quality as research productions.[14]

Stone et al., who differ from Berelson on the extent to which inferences can be made through the use of content analysis, place a great deal of emphasis on categories as integral to both formulation of a study and to its results.

> Our definition presents content analysis as a research tool to be used by the social scientist in making inferences; what is measured in content analysis depends on the theory being investigated. . . . many

---

*Edwin S. Schneidman identified and sought a whole collection of patterns of arguments—logical procedures, rhetorical devices, and psychological mechanisms—in a 1961 study of people's speech and writing. In this study he thought that subcategories were necessary to develop argument pattern measures.

content analysts, often preoccupied with measurement, have felt that they should stay at the level of fact and let the reader draw the conclusions. Actually content analysts invariably use at least a rudimentary theoretical framework in the very design of categories and rules for their application. Its rationale, purpose and implications when applied to the data should be made explicit.[15]

Those categories that have been incorporated into content analysis generally have been of two types: *what* is said and the form information takes, and *how* it is said. Specific categories have included subject matter, direction (the pro or con treatment of subject matter), standards (not usually the researcher's but accepted standards such as balance or fairness or others that have been previously established, values (for example, religiosity or altruism, methods, traits, actors or characters, authority (sometimes referred to as the source of information or opinions), origin of subject matter, target (sometimes expressed implicitly in text or apparent because it appears in items with a clearly defined image), form or type of communication (magazine, newspaper, radio, drama, and so on), form of statement (Lasswell classified this category as threefold—fact statements, preference statements, and identification statements), intensity (strength, excitement value, emotional tone, and so on), and device (often the use of rhetorical devices).[16]

In addition to selecting appropriate categories, the researcher must decide how to utilize them. There are several ways of quantifying the categories discovered in content analysis. Kerlinger writes that there are at least three: nominal measurement, or simple counting of occurrences of each object that has been assigned to a category; ordinal measurement, or ranking of types or numbers of objects per category or across categories or items; and rating based, for example, on previous observations. (This, Kerlinger warns, is subject to errors of bias or a halo effect producing higher subsequent ratings.) Of the three, nominal measurement is most often employed in content analysis with ranking or rating usually reserved for overviews or concluding evaluations by the content analyst.[17]

One possible by-product of nominal measurement is that it may, by recording frequencies, provide a reasonable index of the intensity of attitude inherent in a text. Pool writes that this seems reasonable for a large class of cases.

[T] he assumption baldly spelled out sounds absurd, because it is perfectly clear that frequency is only one of a variety of devices by which feeling is expressed. But the experience of more than one analyst who has tried refinements in measuring intensity has been that nothing much is added by other measures than the frequency one.[18]

Technical Problems

The seemingly straightforward procedure of content analysis can be fraught with technical problems. A great many of these are organizational problems that can be avoided by careful planning and consideration of all contingencies.

One problem area is sampling textual materials. The purpose of sampling is to achieve a representative collection of textual materials. The researcher must consider whether the types of publications chosen, their geographical spread, length of appearance, circulation figures, and other variables produce representative text. After selecting appropriate publication titles, the analyst must select those issues (or items or whatever) that contain pertinent information.[19] Scanning, followed by pretesting, is a good way to winnow out irrelevant material. Some of the questions pretesting should resolve include: Has sampling produced text that applies to the analyst's theoretical framework? Is the category vocabulary adequate? Will this tabulation system enable the researcher to gain needed information without risking tabulation errors of omission or repetition? Are there any ways to economize on the procedure so that only the most germane information emerges?

Other technical problems include validity (accuracy of the measurement vis-a-vis the problem or hypothesis) and reliability (dependability and replicability of the data). Many content analyses do not incorporate measures of the effectiveness of the measuring instrument itself. In these instances, careful attention to sampling, categories, counting procedure, and other details is especially important in reducing the probability of error.

A final matter, alluded to earlier, is that of inference. There is disagreement in the content analysis literature as to the extent to which inference or extrapolation from the data is possible or desirable. Berlson wrote in the 1954 edition of the *Handbook of Social Psychology* that two major problems can arise in content analysis: not applying the techniques of analysis properly, resulting in an incorrect description of the content; or not applying the techniques of interpretation or inference to the correct content description. He held that:

> In a great many studies there is no real problem of inference at all. This is true for all those content analyses in which the description of content itself is the primary objective. Such studies can be said to contain implicit inferences about the causes or the consequences of the content—and some contain them explicitly—but such inferences are in the nature of addenda to or reformulations of the basic data. . . .[20]

The extent to which inference is possible becomes more relaxed in later editions of the *Handbook*. By the mid-1960s it had accepted content analysis as a definite means of making inferences.* As many content analysts now argue, the social context in which communication materials are invariably couched creates something of a mandate for inference making. Lasswell et al. have suggested that some inferences are necessary if only because language itself is so elusive.

> There is as yet no good theory of symbolic communication by which to predict how given values, attitudes or ideologies will be expressed in manifest symbols. . . . There is almost no theory of language which predicts the specific words one will emit in the course of expressing the contents of his thoughts. Theories in philosophy or in the sociology of knowledge sometimes are used to predict ideas that will be expressed by persons with certain other ideas or social characteristics. But little thought has been given to predicting the specific words in which these ideas will be cloaked. The content analyst, therefore, does not know what to expect.[21]

Although the content analyst may not know precisely what to expect, Stone et al. argue that preliminary work can prepare the researcher to create a formal analytic procedure that will serve as the basis for drawing inferences.

> Prior to making a formal content analysis, the investigator can and should carefully inspect a sample of his data, drawing on his intuitive powers to identify the unforeseen circumstances that might be affecting the data. This inspection is used in developing categories and application rules. Formal content analysis is then applied to describe the data as a whole.[22]

These same authors stress the utility of latent as well as manifest implications of content analysis data. While they agree with Berelson that the content analysis procedure itself can only identify manifest symbols or characteristics, they look to an entire body of data or to larger parts of the data to supply a context for interpreting symbols. This arises from their emphasis on the importance of integrating theory with method in the earliest possible stages and throughout an analysis. However, they admit that different researchers can draw different

---

*Stone et al. developed their definition of content analysis (which includes the inference-making function) jointly with Ole R. Holsti of Stanford University in conjunction with his chapter on content analysis for the revised *Handbook of Social Psychology* in 1966. Dr. Holsti subsequently expanded his presentation of content analysis elements and applications in a comprehensive volume entitled *Content Analysis for the Social Sciences and the Humanities* (Reading, Mass.: Addison-Wesley, 1969).

inferences from identical content measures. To adjust for this they offer the researcher a general rule: ". . . the more latent the phenomena inferred, the more internal consistency among categories and external evidence the investigator will want for its support."[23]

In summary, content analysis is an orderly method of observation that strives to provide an objective and accurate measure of the facts and implications of a body of content. Prior familiarity with the content and its social context along with the formulation of a sound theoretical framework are prerequisites to a viable content analysis. Adequate sampling and the selection of essential categories followed by careful recording will facilitate the final step—interpretation of the data. It is at this juncture that the investigator will feel either the confusion or confidence that advance groundwork has instilled. As Berelson observed, "Content analysis, as a method, has no magical qualities—you rarely get out of it more than you put in, and sometimes you get less. In the last analysis, there is no substitute for a good idea."[24]

CONTENT ANALYSIS APPLIED TO THIS STUDY

Guidelines for the Study

Suspicion of monopoly practices or other abuses by major oil companies had been brewing prior to the 1973-74 OAPEC embargo. However, most media presentations of such practices or of the current divestiture debate can be dated from about mid-1973 to the present. Thus, this study looks at media coverage of the issue from March 1973 (six months before the embargo's start) through March 1976 (about the time specific divestiture legislation took shape).

The target for this study is the news and opinion branch of the periodical press—a wide-ranging group of publications that focus on news, business, politics, economics, education, science, and national and international events, using either a news or analytic format or a combined format. Some of these periodicals are limited to certain subjects—this would be especially true of business publications—or limited in their philosophical or political perspectives.* Others concern themselves with the timeliness of what they present while others seek to provide in-depth examination of both the "hottest" and relatively obscure topics. What they all have in common is a commitment to recording the course of events, as they see it, in public affairs.

---

*In this study, business publications, despite their obvious appeal to a unique or self-selective audience, are treated as news and opinion periodicals. While they may be assumed to contain some inherent biases, there is no reason to suspect that these are any greater than in some of the politically oriented periodicals or in those less oriented toward economics.

There are several distinct reasons for choosing this branch of the periodical press for this study. First, print media, because they are static and readily accessible, offer the opportunity to conduct an accurate and systematic examination of an issue over a relatively long time period. Second, periodical publications of national scope and readership are more likely than local or regional print media to contain articles about oil companies, monopoly, divestiture, and related topics.* Third, a study of periodical literature is more manageable than, say, a daily newspaper study of comparable breadth. Fourth, and of prime importance, the news and opinion branch of the periodical press seems most likely of all the media forms to be able to cover this issue in depth and on some sort of sustained basis.

A fifth, more remote, reason for looking at this branch of the press is to be able to infer some of the effects the divestiture issue has had on the general public. Because this segment of the press enjoys a wide-ranging and reasonably well-informed readership, it holds the potential for influencing public opinion. The amount of influence a periodical can actually exert is, of course, dependent on many variables—the topics a publication focuses upon, the extent to which a publication has a definite political image and how that image corresponds with a reader's, its particular editorial policies or approach, its circulation and frequency of publication, its rank in a reader's "library" of periodical reading, the personal influence a reader might have in a community or group (that is, the reader's "opinion leader" status), and so on.[25] Whatever the extent of such influence, these periodicals provide much of the factual and philosophical "grist" for transfer to those who are less informed or whose information is normally gained from other, possibly less thorough, media sources. Consequently, the substance and tone they impart can be viewed as something of a national two-way channel, both reflecting and determining public concepts of the monopoly/divestiture issue.

The decision not to restrict this study either to news or to opinion periodicals was an important one offering several advantages. First, many publications combine news and opinion matters as part of their regular format, so to seize upon either type of publication would exclude those most apt to deal with oil industry behavior or divestiture. Second, the inclusion of both types of periodicals provides a more realistic social context for the issue. Because those who read periodical literature of national scope tend to be exposed to other media and

---

*There are exceptions to this assumption. Newspapers like the New York *Times* and Washington *Post* devote much space to issues of national concern. The *Wall Street Journal*, while technically not a periodical, ran an entire series plus related items on the monopoly/divestiture issue over several months' time. Generally, however, this study assumes that newspapers or the localized press have not assigned much ongoing importance to the divestiture topic unless it is a matter of regional or reader concern.

interchanges with other people, they tend to receive and transmit aggregate or multiple-source messages. So, rather than distinguish and identify these separate, subtle messages, this study attempts to encompass (within the limits of periodical literature) all of them. Third, a look at both types of periodicals is useful in evaluating which periodicals have been most receptive to presenting conflicting information, which have been most factually oriented, which biased, and so on. In other words, a comprehensive comparison of periodicals affecting public perceptions of the monopoly/divestiture issue is a major object of this study. Fourth, an even more central object of this study is to compare and contrast media presentations with the opinions expressed by experts on oil industry structure and behavior. Both news and opinion publications incorporate ideas and information useful in making such comparisons or contrasts.

Sampling Procedure

A review of the *Reader's Guide to Periodical Literature* from March 1973 through March 1976 revealed approximately 300 titles relating to divestiture or allied topics: oil company abuses, responsibility for shortages or high prices, investigations of allegations against the oil industry, specific divestiture proposals, measures to force cooperation or disclosure of information by oil companies, analogies to the old Standard Oil Trust, public attitudes toward oil companies or the energy crisis, and so on. These titles were chosen from a number of headings and subheadings in order to assure that applicable titles were not overlooked: Big Business, Competition, Corporations (Antitrust, Divestiture, Oil Companies, and other subheadings), Monopoly, Oil, Oligopolies, Organization of Petroleum Exporting Countries, and Petroleum Industry.

The titles discarded from these headings included those that did not relate to business in general, the petroleum industry, major oil companies, monopoly, anticompetition or abuse charges, or divestiture. Some titles were included that related to business in general in order to catch spillover commentary that mentions oil company practices and behavior even though it might not treat these subjects as central to the article. This harks back to the notion that the media provide aggregate messages from indirect as well as direct sources.

The initial screening produced approximately 180 titles from 30 publications. Those publications that were eliminated from this list were any that had run fewer than three articles on divestiture or related topics in the 1973-76 period or those deemed inappropriate. "Inappropriate" applied to scientific or academic reviews or journals (for example, *Foreign Affairs*, *Annals of the American Academy of Political Science*), reprints (for example, *Vital Speeches of the Day*), or those that did not fit into the news or opinion category for a broad class of readers (such as *Motor Trend*, the *Department of State Bulletin*).* This

---

*Those publications deemed inappropriate also averaged only two articles per publication, placing them below the frequency as well as the appropriateness criterion.

reduced the sample to 11 publications and approximately 150 articles. A scan of a sample of each periodical's articles led to the construction of the content analysis elements and categories and provided the form for a pretest of these categories.

Element and Category Construction

The major element or unit of the study is the periodical as a whole, followed by the item or individual article—its subject matter (does it deal directly with divestiture or with related topics?), its direction (is it pro, con, or neutral on monopoly charges, divestiture, or other actions?), and its intensity (what is the tone or depth of conviction being applied to the central ideas expressed in the article?). These three aspects of the item might be said to correspond roughly with Osgood et al.'s Potency, Activity, and Evaluative dimensions of meaning.[26] In this study, items will be examined to see whether particular subjects are predominant, whether subjects are treated neutrally and passively or with a definite perspective and active stance, and whether value judgments are made about certain kinds of subject matter.

The item, in this case, however, provides only a crude measure of the ways news and opinion publications have presented this issue. Because a number of conflicting ideas and arguments on the subject have been expounded by a variety of people, three additional measures are necessary: the types of arguments presented, the sources of arguments or information, and the context of the presentation or type of article. Length, direction, and frequency measures are applied to these categories to obtain greater information about the extent or intensity of certain categories and items.

For example, a news item (type of article) has as its subject matter "Senate efforts to get oil company executives to disclose industry data." Although this does not deal directly with monopoly charges or divestiture per se, it may incorporate statements by senators to the effect that the oil companies have colluded with one another to keep production figures secret, a type of argument that appears in more pertinent articles. The frequency (and relative space) of these senators' remarks are noted, but the article itself is not recorded as having a particular direction or intensity unless the author of the article (also, at times, a form of source) makes judgmental comments such as, "The Senators' indignation was fanned by the corporate heads' united lack of cooperation." This fictitious article might be recorded as: news; ½ page; author unnamed; subject—disclosure of industry facts withheld; sources—Senators Jones, Smith (twice), and James (anti-major-oil-companies) and corporate heads Carnegie and Morgan (pro-industry); types of arguments—collusion (3 times), secretiveness (4 times), insensitivity to public needs (2 times), scapegoating (2 times), badgering corporate heads (2 times); direction—somewhat anti-industry; comments—notation of specific quotations that express the sssence of the article. A totaling of the arguments and sources used, combined with the author's comments might be summed up as a news article of moderate anti-oil-industry bias focusing on problems of disclosure. The recurrence of such articles could eventually form a pattern for evaluating an entire publication's editorial position on divestiture, monopolistic or anticompetitive behavior by oil companies, or related topics.*

In this study, a number of categories comprised the content analysis pretest. Once these had been applied to two or more articles per publication, a final test list of categories emerged. This list was further enlarged in a few category areas during the actual analysis as a few new arguments, subject matter areas, types of articles, and so on appeared. Also, a few subcategory areas, though retained, proved only minimally useful. The basic pretest and test incorporated the following variables (italicized items were added as a result of the pretest):

1. Name of publication
2. Date, page numbers, length of article
3. Author (if any)
4. Title
5. Type of article:
   a. editorial
   b. analysis
   c. news
   d. *opinion analysis* (for example, outside contributor)
   e. interview
   f. *news/analysis* (analysis relating to specific event)
   g. other
6. Subject matter
   a. divestiture
   b. partially divestiture
   c. monopoly/anticompetitive behavior
   d. high prices, shortages
   e. *oil company abuses*
   f. *oil company image*
   g. *disclosure*
   h. excess profits, reporting profits
   i. government's role vis-a-vis oil companies
   j. other

---

*An example of a related topic might be the effects of high prices on consumers and the economy that (the article hints) stem from major oil companies' control of the means of production. Although not directly termed a monopoly or divestiture topic, this actually says a great deal about the concept of the issue held by the author(s) or editor(s) of such an article.

7. Kinds of evidence or arguments used (note whether pro, con, or neutral toward oil industry or major oil companies; list new arguments, count frequencies)
   a. market share data
   b. ease of entry of competitors data
   c. conduct data
      (1) cooperation or collusion with other oil companies
      (2) *cooperation with OPEC*
      (3) squeezing independents off from supplies
      (4) *withholding supplies*
      (5) *advertising or marketing practices*
      (6) *failure to disclose data*
      (7) price gouging
      (8) *price cutting*
      (9) joint ventures
      (10) *scapegoating oil companies*
      (11) *trying to improve industry image/relations*
      (12) other
   d. profits
   e. other performance measures (list)
   f. integration (unspecified)
   g. integration (vertical)
   h. integration (horizontal)
   i. responsibility for high prices, shortages, or crisis
      (1) OPEC
      (2) oil companies
      (3) major oil companies
      (4) multinational oil companies
      (5) government policies (list)
      (6) other
8. Sources cited (list and count frequency for each; indicate whether pro, con,
or neutral toward industry, majors, independent, government action, etc.)
9. *More space in article is devoted to sources whose stance is*:
   a. *pro-oil-industry*
   b. *promajor*
   c. *anti-oil-industry*
   d. *antimajor*
   e. *neutral*
   f. *other* (for example, pro-big-business, pro-independent but not anti-
      major, and so on.)
10. Attitudes/opinions reflected in article (note whether very favorable/very
unfavorable, favorable/unfavorable, only partially or mildly favorable/un-
favorable, or neutral)

a. business in general   )   (These three categories were least
b. big business   }   useful but served to provide con-
c. small companies   )   text for some articles.)
d. major or large oil companies
e. independent oil companies
f. market mechanism (allowing market to set prices, for example)
g. government action (specify which measure)
h. oil industry
i. competitiveness of oil industry
j. divestiture

11. Quotations or comments from article that express its central thought or
ideas or relate something of the tone of presentation.

Certain premises or decisions guided the construction of the preceding
categories. First, because the study is especially concerned with expert versus
public, government, or media perceptions of the competitive nature of the oil
industry, source is an essential and central category. Sources can be people or,
less often, groups (both named and unnamed), or published materials, statistics,
reports, surveys, and so on. The way a source is used is especially important. For
example, a top Treasury Department official might normally oppose divestiture,
regarding it as anticompetitive or deleterious to trade. But, if this source is
quoted as saying he feels the FTC investigation is worthwhile, readers might assume that Treasury would support the breakup of vertically integrated firms.
The source may have been taken out of context. If sources have been used to
support views not necessarily their won, the content analysis takes note of this
in the comments or perhaps the attitudes/opinions-reflected categories. However, the source category is used solely to record what the source was and how
that source was used, not whether the source was ill used. This recording decision is based on the assumption that a writer, if trying to make certain points,
will be more apt to choose sources whose views are compatible with, not adver-
saries of, these points. The analysis revealed some exceptions to this assumption, most notably in articles designed more to persuade than inform. But
writers did not appear to try very often to trap sources in their own words.

One thing the pretest did suggest was that another category should be
added to weight the effects of source. Category 8, the source category, counts
the number of references to a source and notes whether each source, as used,
presents a particular viewpoint (pro, anti, or neutral toward the oil industry, for
example). Category 9 calls for an informal notation as to which sources were
given the most physical space in an article. This was done to get a finer tuning on
how sources were used to create different moods or emphases. For example, one
article that carefully balanced the number of sources with opposing views
quoted government sources at some length. When presenting the industry's
view, the author was sparer in detail, quoting one source as having said merely,

"Baloney." This category, admittedly only a rough estimate of proportionate source space, still provides some additional information in evaluating an article's or publication's coverage of an issue.

Equally important is the type of argument or evidence employed to illustrate certain ideas or opinions. Unlike Berelson or some of the other content analysts discussed in the previous section, this study does not take pains to specify the form an assertion, theme, device—considered here as a type of argument—must take. Rather, it is concerned with the substance or origin of the argument that singly, or in conjunction with other arguments, leads to some sort of conclusion: there is evidence of monopoly, there is not evidence of monopoly, there may be, it is inconclusive, divestiture will restore competition, divestiture is a poor solution, and so on. While trying to identify arguments as to type and direction, this category's function is not to determine the tone of the article. Instead it tries to record objectively what rationale comprises a thought and how often it is brought to bear on an issue. If an argument uses epithets or strong language, some note may be made of its tone; but with rare exceptions, other categories (attitudes/opinions reflected, and comments) are used to assess the tone, slant, or bias of an article.

The subcategories used in the arguments category were gleaned from the media, public records, and academic literature. Although a few additions proved necessary in the actual content analysis, these basic arguments were quite adequate for most of the publications studied. The arguments were chosen to achieve several objectives. One objective was to test which arguments are most popular with which publications and with the news and opinion periodical press as a whole. A second objective was to examine the different ways these arguments are used. For example, in the actual content analysis it was not uncommon to find two philosophically disparate periodicals using the same argument to demonstrate opposing points of view. Market share data was one of these conveniently twisty "objective facts." One publication might decry the concentration of power produced by the top eight firms controlling 53 percent of refining capacity in 1974. Another publication might point out that this figure represents a drop in refining capacity relative to independent companies' rising share from 1970 to 1974. Likewise, profits may be reported as excessive or windfall by one publication, whereas another may rejoinder that a flagging domestic industry now has proper investment incentive. Thus, this study takes note of in whose favor the arguments fall.

A third objective of the arguments category was to test the hypothesis that traditional indexes favored by many economic experts on the oil industry (such as market share, ease of entry, and firm conduct measures) have taken a back seat—in media presentations—to implicit-behavior arguments such as squeezing of independents, unfair marketing or price discrimination, special deals (for example, joint ventures) and other quasi-legal sidesteps. The media's choice of facts and arguments, if they seem consistent over time, should enable some inferences about the media's information-gathering strengths or limitations, and possibly about its own persuasions on the matter.

The subject matter chosen for an article coupled with the type of article are two additionally useful categories. They indicate the extent to which a publication considers topics important. Further, they indicate whether an editor or author considered the subject newsworthy, deserving of special consideration (an editorial, special feature, interview or the like), requiring in-depth treatment (analysis, news/analysis or a recent event or opinion/analysis by a columnist or outside contributor), or some combination of the preceding. For example, if a publication persistently focuses upon the woeful effects of higher prices, it indicates a concern that remedies be sought and applied (this may or may not entail a stance on who is to blame for higher prices). If a publication repeatedly states its opposition to government intervention in general, it is not likely to support price controls or excess profit taxes for the oil industry in particular. If a publication often features interviews with independent oil producers, it may be thought of as biased against major, vertically integrated firms. However, all of these contentions require support. One category will supply some of the facts, but inferences can be drawn only if combined categories reveal a pattern, force, or direction for a particular article or publication. For example, the publication interviewing independent producers may reveal, through its editorial and analysis articles, that it supports the oil industry's position on price decontrols and that it, in fact, is not biased against large firms, but does have a penchant for the "frontiersmen" image of the wildcatters.

Another category, attitudes/opinions reflected, is among the most subjective. In this study, the terms attitudes or opinions are used rather loosely and interchangeably to refer to the biases, leanings, or general direction of thought (derived as much from covert statements as possible) applied to several concepts or entities—big business, small businesses, big oil companies, independents, government action, market forces, the oil industry, divestiture, and possibly a few others. Fortunately, some of the other categories—notably, type of article, source, and types of arguments—assume the most rigorous and most easily quantifiable role in the analysis. The attitudes/opinions category is used in a supplemental way, to enhance interpretation of meaning or to show whether an article has been judgmental or has used language that posits a definite stance. This category is called into play primarily for editorial, analysis, or opinion/analysis articles. This often happens automatically because most news, news/analysis, and many analysis articles will simply not provide a forum for opinion; thus they are accorded a neutral score. Further, where the attitudes/opinions subcategories received favorable or unfavorable marks (on a seven-point continuum), special efforts were made to supply comments (Category 11) to support this rating and to illustrate the direction or intensity of the article's views.*

---

*In order to note direction, a rating scale of 7 points on a continuum was used: very favorable (VF), favorable (F), partially or mildly favorable (F-N), neutral (N), partially or

The final category, comments, or the selection of quotes, is also a highly subjective category—one that could have almost as many vairables as there were researchers. Its purpose is to focus on sentences, phrases, or passages that convey the central ideas and/or tone of an article. In this sense, if there is a pivotal thought or expression in the article, many researchers might be tempted to seize upon many of the same quotes. (For example, the fact that the United States, during one week in 1976 and for the first time in its history, imported more than half the oil it needed was a fact repeated excitedly by many publications.) Ultimately, of course, the reliability of this particular category will depend upon research beyond the scope of this study. Suffice it to say that caution and attention to all the other categories should temper the selection of direct quotations and the interpretation they receive.

Summary of Sampling and Pretest Results

The pretest helped in developing a number of categories and subcategories to measure factual content, test various hypotheses, and identify certain patterns of direction and intensity in articles on oil industry structure and behavior and/or divestiture. A rough screening of approximately 150 articles reduced the number of relevant articles to 127. To afford greater context for interpretation of the data, many articles were considered that dealt only in part with topics of monopoly, competition, or divestiture. (For example, some dealt with prices, some with government's role vis-a-vis the oil industry, some with shortages, and so on.) Thus, the figure 127 may be a bit misleading. This will be dealt with in the next three chapters, which report on the actual content analysis.

In sum, 11 news and opinion periodicals published 127 articles that touched upon this study's subject, during the period March 1973 through March 1976, as follows: *Business Week*, 23 articles; *Dun's Review*, 3; *Forbes*, 8; *Fortune*, 5; *The Nation*, 10; *National Review*, 9; *The New Republic*, 19; *Newsweek* 23; New York *Times Magazine*, 6; *Time*, 11; *U.S. News & World Report*, 10.

CONTENT ANALYSIS OVERVIEW

The content analyses that follow are grouped according to type of publication: business periodicals (Chapter 4), news magazines (Chapter 5) and news/

---

mildly unfavorable (N-U), unfavorable (U), and very unfavorable (VU). There was also an eighth rating scale for inapplicable responses. These ratings were devised to aid in evaluating explicitly attitudes or opinions in articles or to note that neutrality had been the standard.

opinion publications (Chapter 6). A seventh chapter will explore the explicit attitudes of journalists and economists as expressed in their responses to a questionnaire. A final chapter will address itself to comparing and interpreting the content analysis and questionnaire data and will evaluate the performance of the news-opinion branch of the periodical press in presenting the oil monopoly/divestiture issue.

NOTES

1. Bernard Berelson, "Content Analysis," in *Handbook of Social Psychology*, ed. Gardner Lindzey (Reading, Mass.: Addison-Wesley, 1954), p. 489.

2. Fred N. Kerlinger, ed., *Foundations of Behavioral Research* (New York: Holt, Rinehart and Winston, 1967), p. 544.

3. Ibid.

4. Phillip J. Stone et al., *The General Inquirer: A Computer Approach to Content Analysis* (Cambridge, Mass.: M.I.T. Press, 1966), p. 5.

5. Ibid., p. 17.

6. Harold D. Lasswell, Daniel Lerner, and Ithiel de Sola Pool, *The Comparative Study of Symbols* (Stanford: Stanford University Press, 1952), p. 65.

7. D. Cartwright, "Analysis of Qualitative Material," in *Research Methods in the Behavioral Sciences*, ed. Leon Festinger and Donald Katz (New York: Holt, Rinehart and Winston, 1953), p. 447.

8. Stone et al., op. cit., p. 7.

9. Berelson, op. cit., p. 508.

10. Stone et al., op. cit., pp. 35-36.

11. Berelson, op. cit., p. 509.

12. Berelson, op. cit., p. 511.

13. Berelson, op. cit., p. 509.

14. Ibid., p. 510.

15. Stone et al., op. cit., p. 14.

16. Berelson, op. cit., pp. 510-11.

17. Kerlinger, op. cit., p. 550.

18. Ithiel de Sola Pool, "Trends in Content Analysis, A Summary," in *Trends in Content Analysis*, ed. Ithiel de Sola Pool (Urbana: University of Illinois Press, 1959), p. 194.

19. Berelson, op. cit., pp. 515-16.

20. Ibid., pp. 516-18.

21. Lasswell, Lerner, and Pool, op. cit., p. 49, cited by Stone et. al, *General Inquirer*, op. cit., p. 10.

22. Stone et al., op. cit., pp. 16-17.

23. Ibid., p. 17.

24. Berelson, op. cit., p. 518.

25. See Melvin L. DeFleur's discussion of the social relationships theory of communication in Melvin L. DeFleur, *Theories of Mass Communication* (New York: David McKay, 1966), pp. 129-33. Certainly some newspeople have earned the reputation of nationally based opinion leaders, affecting national audiences who perceive themselves as in agreement with and influenced by, for example, Morley Safer or George F. Will.

26. C. Osgood, G. Suci, and P. Tannenbaum, *The Measurement of Meaning* (Urbana: University of Illinois Press, 1957). For an abbreviated discussion of semantic differential and dimensions of meaning see Kerlinger, op. cit., pp. 564-80.

4

THE BUSINESS PERIODICALS:
A CONTENT ANALYSIS

BUSINESS WEEK

From June 1973 through December 1975, this weekly business magazine featured 23 articles or a total space of approximately 29 pages on divestiture, oil industry competitiveness, and related topics. More than half (14) of these articles appeared in 1974. Most of the articles were news articles (11). There were also four editorials, three news-analyses (in-depth articles of a topical nature), two regular analyses, and one interview, one personality profile, and one book review. This book review was the only article that credited an author—in this case, the publication's own energy editor. (However, issues beyond the March 1976 period of study show a slight shift toward more by-lines.)

Subject matter for these articles was diverse, tackling and exploring in something of a devil's advocate fashion the problems that confront the oil industry. Although none of these articles dealt solely with divestiture, six of them focused a great deal of attention on the subject. Two additional articles dealt with "forced divestiture," a somewhat misleading term that, in fact, means voluntary divestiture or selling off of operations to avoid or deter financial losses. More than 17 of the 23 articles touched on or discussed at length the cloud hanging over the oil industry as a result of criticisms made by public officials and private citizens (insufficient competition, lack of disclosure of information, special arrangements among majors), the FTC investigations and antitrust lawsuits, public opinion polls on who is to blame for oil shortages and high prices, the poor image of oil companies, and independent oil companies' criticisms of the majors. Such concerns were illustrated by such titles as "An Ominous Mood for Oilmen," "Legislatures Make Trouble for Big Oil," "Oil Companies Get Hit by a

Poll," "Oil Lobbyists Face a Tougher Audience," and so on. The remainder of the articles dealt with the dilemmas of counteracting government criticisms or riding out the storms of government regulations (for example, Federal Energy Office fuel allocation programs or price controls), while at the same time trying to bargain with OPEC and find means of investment in order to expand the domestic petroleum industry. Some common complaints voiced by two sources with disparate views illustrate the tension that appears in many *Business Week* articles:

> "What we object to most is the 'save the small businessman' routine that keeps coming out of Washington," says Owen L. Hill, president of the Clark Oil and Refining Corp. "Government regulations penalize companies that are most aggressive in taking care of themselves."

and:

> The two-tier pricing system "is forcing the major oil companies to behave like monopolists, undercutting the prices of their smaller, independent competitors, according to an unpublished study prepared by the FEO.[1]

In addition, three articles devoted space to a careful explanation of what windfall profits actually meant in terms of the investment profit rates of other industries, what effects these would have on the general economy and the industry, and how these came about. One editorial, for example, although defending industry in part, accepts some penalties:

> The rush to tax windfall profits . . . would certainly have resulted in stultifying domestic investment by the oil companies just when the country so desperately needs gargantuan expenditures to step up exploration. . . .
>
> There are better and more sophisticated responses to excessive oil profits, responses that recognize the need for investment, that retain incentives to produce and market increasing volumes of oil products, and that take note of the character of the oil industry. The big international oil companies, after all, no longer control the game.[2]

Instead, the editorial recommends an end to the depletion allowance and to the preferential tax treatment that allows companies to treat royalty payments to producing countries as taxes already paid to the U.S. government, along with a rollback in the "prices of products grossly affected by increases in the cost of offshore crude. The oil companies ought to absorb part of these costs themselves."[3]

Throughout the articles there is a somewhat defensive approach to subject matter selection—seize the thorniest industry problems as soon as any new developments arise, present them point-blank, then get sources from many angles to illustrate the threat or opportunity this poses to the industry. Interestingly, this "let's face facts before they face us" emphasis, coupled with a strong news format, produces a number of arguments and sources that are directed against the oil industry, especially the major integrated firms. The arguments used are largely those implicit-behavior arguments favored by the FTC and prodivestiture senators. For example, the argument that the large, vertically integrated companies effectively control supply and are thus able to squeeze independents at refining and marketing levels was mentioned four times. In one of these cases, this argument was used to combat price control and allocation formulas thought by independents to leave the major oil companies with most of the low-cost crude and continue the "competitive disadvantage of the independents."[4] Only one argument was presented that denied squeezing had taken place. Articles made references seven times to "sweetheart" arrangements or cooperation among majors through sharing refined products or oil, joint ventures, price fixing agreements with OPEC or with other majors, and obstructing independents at a marketing level. Only two arguments viewed sharing or joint ventures as procompetitive.

One argument that was not anti-industry concerned the profits area. Arguments were essentially balanced, tipping slightly toward the profit makers. Five arguments claimed that profits were excessive, whereas six viewed profits as a result of higher world oil prices or temporary inventory profits and not excessive in view of previously lagging industry performance. For example, one interview quotes Mobil Oil Chairman Rawleigh Warner, Jr. as saying:

> If this industry is going to do all of the things the consumers of the world want to do, we have got to have the income to borrow the money—and pay it back—to do this job. People are beginning to accuse us of enjoying windfall profits when in point of fact we are not going to enjoy them. It's for the simple reason that while prices at this moment outreach costs, costs are going to go up.[5]

Another set of arguments criticized industry failure to produce data requested by investigatory committees and accounting methods on five occasions. However, two articles incorporate defenses of certain bookkeeping practices, in one case stating, "Some analysts and critics grumble that reserves appear at a particularly 'convenient' time. But setting aside such funds is largely a matter of company and accounting discretion."[6] And the problem of poor industry image, presented on three separate occasions, was treated sympathetically and was viewed as a situation requiring prompt and remedial action, as illustrated in the following excerpts:

A dozen Congressional committees are investigating the industry, including Jackson's permanent investigations subcommittee. Last week's fiery session was only the beginning of a much deeper probe that committee staffers hope will rival some of the committee's hearing on labor racketeering. To some oilmen, the accusatory manner of questioning last week bore a resemblance to the racket hearings. "For God's sake," cries Mobil's Warner, "we're being treated like criminals."[7]

. . . Meanwhile, a far more serious threat is taking shape: a plan to break the majors up into separate producing, refining, transporting and marketing companies . . . to fend off such challenges, the oil companies are working fast to improve what has overnight become the worst image in American business. . . .

A public-be-damned attitude was fostered by the industry's close ties with such one-time powers in Congress as Sam Rayburn, Lyndon Johnson and Robert Kerr, all from oil-producing states. "We were lulled into a false sense of security in those days." Continental's Hardesty explains, "because we knew that the leaders in Congress understood our business. We reacted far too slowly when Congress changed and all of a sudden we had to deal with people who didn't understand us."[8]

In recent weeks the public's attitude toward the energy crisis has grown sharply skeptical, and most of the suspicion that the whole thing was contrived to raise prices has fallen not on the Arab nations and their embargo but on U.S. oil companies and their industry. . . . The public's casting of the petroleum companies in the villain's role is a major finding of a poll by the Opinion Research Corp. . . . This attitude, ORC warns, could have broad consequences, possibly creating "a backlash against all business" in "today's adverse business climate."[9]

Those arguments that assign responsibility for high prices, oil shortages, or the energy crisis in general also take the form of industry criticisms. Oil companies or major oil companies are blamed 21 times for prices or shortages, whereas OPEC takes the brunt of such blame only 8 times. However, in editorials and analysis articles, *Business Week* clearly points to OPEC as the cause of price discrimination and energy gaps:

The Ford Administration struck a new and welcome note of realism this week with its attack on the   exorbitant price policies of the oil exporting countries. For the first time, it appears, U.S. policymakers have taken a hard look at what the oil cartel is doing to the world economy. . . .[10]

Subject to some criticisms as well were government policies—partially the fault of government itself but also taken advantage of by industry—such as the depletion allowance (three times), foreign tax credits (two times) and government laxity in regulating or monitoring oil industry activities (five times). Mentions of price controls, however, were always used to express industry dismay with such measures. A final argument—the extent to which vertical integration inhibits competition—was presented five times in statements opposing vertical integration and only twice in its favor.

In total, the *Business Week* articles incorporated 93 argument references. Seven of these decreid government policies, 27 supported the major oil companies' contention that they behave competitively and fairly, and 59 arguments were of an anti-major-oil-company nature ranging from squeezing to price gouging to vertical integration.

Business Week's use of sources is more revealing of certain biases than its presentation of various arguments, though even here there is not the clear pro-industry perspective one might expect of a business publication. Of a total of 142 sources 66 were connected with the oil industry, 50 were government sources, 22 were nongovernmental critics of the oil industry, 4 were independent consultants or opinion polls (there were no academic consultants or experts mentioned), and the remainder were impartial, neutral, or incidental sources.

Of these considered industry sources, 33 represented major oil firms, 6 were from middle-sized integrated companies, 16 were independents, and 11 were professional oil industry consultants. The most frequently cited sources (those cited 3 or more times) were Rawleigh Warner, Jr., Mobil chairman and former American Petroleum Institute chairman (10 times); Anthony Sampson, author of *The Seven Sisters*, a book not favorably disposed toward the major multinational oil companies (7 times); J. K. Jamieson, Exxon chairman (5 times); and Herman M. Frietsch, official of a U.S. oil equipment supply firm accused of colluding with Venuzuelan oil producers to restrict supplies (3 times). The source most often referred to (mentioned 7 times but not once actually cited) was Senator Henry M. Jackson. In fact, senators as a group were the most frequently mentioned (14 times), but were cited directly only 5 times. Next to sources associated with major oil companies (28), industry critics were the most frequently quoted group (17 times). Industry critics were not named on five occasions as was also true for two congressional sources and two industry spokesmen. Also quoted extensively were 16 sources associated with independent oil companies and 11 government officials with assorted cabinet, commission, and administrative posts at both national and state levels.

Another measure, the way sources were actually used in clarifying facts or positions, indicates a careful—perhaps even a self-conscious—attempt to balance sources. Sources whose stated position at the time represented a pro-oil-industry response numbered 45. One source made a specifically promajor statement and 9 sources spoke out specifically against government intervention in the oil marketplace. Thus, the total number of sources who looked favorably upon the oil industry was 55. Conversely, 38 statements expressed disagreement or displeasure with oil industry conduct or structure; 7 were specifically anti-multinational-corporations; 3 were specifically directed against major, vertically integrated firms; and 10 favored government policies that would alter or limit oil company business activities or penalize them—a total of 58 sources with anti-oil-industry sentiments. Fifteen sources were neutral or noncommittal.

In terms of amount of coverage given various sources, a moderately pro-industry position appears. In their relatively generous coverage of the opinions and views of consultants (11 oil or investment consultants), four articles devoted more than half their space to sources appreciative of the oil industry's problems but not keenly affected personally by its ups and downs. In total, two articles were decidely pro-independent, one balanced source space, six gave anti-industry sources more space and ten made pro-industry sources predominant. This 8 to 14 (anti-industry to pro-industry), ratio, while more telling than the number of sources, when looked at in view of other categories, does not seem to indicate an overwhelming bias toward hiding the oil industry's flaws and foibles.

The attitudes and opinions reflected in *Business Week*'s 23 articles tended to follow or parallel the types of articles: 11 news, 4 editorial, 5 analysis or news/analysis, and 3 review/profile stories. Of these, three articles expressed the necessity of supporting the aims of business or industry in general. The remainder stayed neutral on this point. Of course, some might argue, plausibly, that a business publication has little need to state an obvious bias. Nonetheless, there was little overt hornblowing and what there was occurred primarily in editorial material, or in allowing quotations by individuals with strong business biases. Where this publication took the strongest probusiness stand was in its repeated opposition (again, mostly in editorials) to mandatory policies or controls that would hamper free enterprise.* As one editorial criticizing Congress's reluctance to accept natural gas and other fuel price decontrols argues:

> One way or the other, energy prices must rise to spur conservation and stimulate new energy development. . . . Instead of wasting time kicking up their heels, the members of Congress should buckle down to negotiating the compromise policy the country needs.[11]

Of the 23 articles, only one—an interview providing an exclusive forum for an oil executive's views—could be said to expressly favor big oil, although in this

---

*An average of three articles opposed various types of government interference: price controls, fuel allocation programs, a federal energy company, and divestiture. However, two articles favored having government get tough with OPEC and two favored forced divestiture if such a move could be shown to simplify and aid oil industry operations.

case, the interviewer, if anything, challenged the interviewee. Five articles sympathized vaguely with major oil companies and the rest remained neutral. Predictably, none of the articles contained antimajor biases, although one article was devoted to independent sources who voiced many criticisms, and one featured (rather flatteringly) the Federal Trade Commission lawyer who was charging eight majors with monopolisitc behavior.

The oil industry as a whole was depicted favorably (though somewhat chidingly at times) in 12 articles, unfavorably in 1, and in a neutral tone by the remaining 10 articles. Two articles promoted the competitiveness of the oil industry while two other articles critically questioned some of its practices (for example, use of foreign tax credits). One article gave slight praise to the industry's competitiveness and the rest of the articles made either no comment or remained noncommittal.

On the key question of divestiture, surprisingly little comment was made. Two articles suggested that divestiture (but only on a voluntary basis) might not really hurt and might even help some oil companies. One article opposed the FTC and Senate brand of divestiture; the remaining 20 articles indicated no preference. Again, this and other silences may stem from the "kid-glove" tendencies of *Business Week*, its acting as somewhat as an assistant ringside coach—providing towels and water (the basic news material) but going low on pep talks, and letting business do its own slugging.

DUN'S REVIEW

This monthly publication published three articles (in July 1973, April 1974, and February 1975) or a total of 9-1/3 pages on oil industry behavior and antitrust matters. Two of these were interviews, one with a Harvard Law School professor critical of the Federal Trade Commission's anti-trust program, and the other with a political geographer critical of both the OPEC cartel and multinational oil corporations. The third article was an analysis of the oil industry's poor public image. All three were written by publication staff members.

The articles incorporated 17 pro-industry and 15 anti-industry arguments. These included cooperation of multinational companies with OPEC (one confirmed, one denied), alleged withholding of oil supplies until price decontrols occurred, excess profits (not believed to be excessive), OPEC's responsibility for high prices (4 blamed OPEC, 3 blamed oil companies), and the industry's failure to communicate successfully with either press or public.

It was the topic of industry's poor image that caused *Dun's* to incorporate so many arguments critical of the oil industry. These appeared mainly in an article entitled "Oil Meets the Press," in which the industry is largely blamed for its own image problems.

Part of [the] reputation for villainy can be blamed on hasty and ill-informed reporters and editors, part on agile and opportunistic politicians and part on an angry and perplexed public too ready to accept uncritically any and all charges made against the oil industry. But year in and year out, no segment of American society has done more to muddy the public reputation of the oil industry than the oil industry itself.[12]

Complaints of inaccessibility to the press, secretiveness or guardedness in dealing with public officials, and insensitivity to public concerns and needs were the main charges this article brought to bear on the industry, particularly the majors. (In fact, no mention of other than major or multinational companies is made in *Dun's*.) However, the article sounds a faintly optimistic note when it points out that many oil companies have taken their poor image in tow and tried to improve it.

Exxon's Jamieson and President C. C. Garvin, Jr. have been talking themselves hoarse for weeks, opening their doors to everyone from *Time* magazine to the *Nutmegger*, a small New England publication. Day after day they shuttle between the executive suite on the 51st floor and an impromptu studio in the basement, where the networks now come and go routinely to tape interviews.[13]

But the effects of such suddenly responsive efforts, the article concludes, are in doubt and may even reflect badly on business in general.

In the remaining two interview articles, the arguments used—mainly by those being interviewed—criticize the FTC on the one hand (for its handling of all types of antitrust cases, including that against the major oil companies), and the multinational corporations on the other, a relatively balanced set of pro- and anti-industry arguments.[14] One interviewee, an academic lawyer, argues that the FTC's interpretations of the laws against price discrimination have had the effect of protecting only some companies from competition, resulting ultimately in higher prices. He also charges that the FTC, because it requires congressional approval of its budget, is more susceptible to political pressures than, for example, the Justice Department and has also engaged in more frivolous suits than Justice. The other interviewee, an academic political geographer, claims that the multinational oil companies arranged with OPEC to help raise world oil prices (the interviewer was unable to shake him from this assertion). But even here, the predominant blame is viewed as OPEC's rather than industry's.

If the arguments in *Dun's* include some little-discussed concepts (as in the interviews), some of its sources are equally off the beaten path. In addition to the not-so-unusual major oil company executives and public relations people, there were two academic sources, one journalist for a weekly petroleum publication, one sheik, an environmental group, one senator, and an airplane pilot. Of a total of 35 source references, 22 were cited and 12 mentioned. Fourteen of these were pro-industry sources, 13 of them associated with major oil companies and the one press source; 8 were critical of some aspect of government; and 13 criticized or were typified as critical of the oil industry. The only congressional source mentioned (but not cited directly) was Senator Henry Jackson. As for source direction, 14 had a pro-industry, promajor stance; 8 were critical of some aspect of government; 11 were anti-oil-industry; and 1 was neutral. In terms of amount of coverage given different sources, the pro-industry sector predominated in one article, was given some preference in another, and was barely visible in a third.

In the attitudes and opinions category, buisness in general was treated very favorably in one article, favorably in one, and not at all in another. Big business was supported only slightly by one interviewer and small companies or entre-preneurs (including independents) received no notice. Big oil received favorable treatment in one article (though there was a pronounced admonition mixed with the boosterism) and neutral treatment in the other two articles—from the inter-viewers', not the sources' viewpoints, that is. The issue of government interven-tion per se, including divestiture measures, was not touched upon. However, there were some criticisms of particular government measures and what were perceived by one author as political ploys. The press, the public, and the oil in-dustry all received bad marks for misperceiving one another. The competitive-ness of the oil industry lay essentially unquestioned, suggesting perhaps that there is no question in the minds of *Dun's* editors as to how readers should regard this matter.

In sum, *Dun's* gave scanty and rather pro-oil-industry coverage to industry behavior during March 1973 through March 1976. There was an oblique concern with the FTC's role in bringing antitrust suits (a matter of general, not just oil-industry concern), and with OPEC's responsibility for the energy crisis and high prices. But mainly there was a focus on image, a concern that remedial action in repairing big oil's tattered dignity had come too late. As one of *Dun's* editors ex-plained it:

> What the industry had failed to realize was that while their lobbying efforts could work on such esoteric issues as the oil-depletion allow-ance, which neither interested nor discomforted the public, the mood would completely change as soon as the public became en-raged over long lines and skyrocketing prices . . . concedes Raymond D'Argenio, public relations director for Mobil, "and now the general public is mad as hell at us."[15]

FORBES

From February 1974 through January 1976, *Forbes*, a business biweekly, published eight articles or a total of approximately 18 pages on oil monopoly, divestiture, and related topics. None of these bore authors' names although a staff writer was alluded to as author in one article. Six of these articles were analyses (highly opiniated analyses, it should be noted), one was an editorial, and one an interview/analysis.

Topics included divestiture (as the focus of one early 1976 article and discussed, but not the central topic, in two previous articles), profits (two articles), punitive actions against or criticisms of industry (four articles), accounting and reporting procedures of oil companies (one article), OPEC pricing (one article), and the need for government policies that either provide investment incentives or at least avoid creating disincentives to industry (three articles). One entire issue was devoted to big business. In fact, the common thread throughout was the problems business, but especially big business, faces in convincing the public that it has not behaved monopolistically or tyrannically toward either small businesses or consumers. As one article explains it,

> Americans have always been ambivalent about bigness. On the one hand, bigness is good; it's a big country, and you can't have too much of a good thing. But on the other hand, Americans are strong individualists, and in that respect, bigness is bad; it threatens the individual, the small businessman, the local government.[16]

Then it goes on to decry this ambivalence.

> In the economic sphere, our antitrust tradition is a sop to this individualistic antibigness. By and large, it is only that—a gesture. The simple fact is that our society wouldn't be what it is today—for better or worse—without huge enterprises.
>
> At the time the Standard Oil Trust was broken up in 1911, its 34 constituent companies had sales of $95.4 million. Last year the five principal companies carved out of the trust grossed more than $50 billion. . . . Antitrust has scarcely even slowed the trend to bigness. Last year there were 248 companies with sales over $1 billion. Fourteen companies netted more than $500 million; three more than $2 billion. . . .[17]

Another article echoes a similar thought.

> . . . Should Big Oil be broken up to prevent undue concentration of power?. . . Americans have never given a clearcut answer to this question. The Sherman Antitrust Act and the whole line of laws and court decisions that followed in its Jeffersonian train was one response. But in many other areas we permitted concentrations of power for economic efficiency.
>
> In granting a near monopoly on telephone service, for example, we received in return a low-cost product and a fair measure

of responsiveness. . . . We permitted huge labor unions to force their will on fragmented industries like trucking and construction. We have given Big Government increasing power over everyone's lives. Whatever tradition may say, Americans are ambivalent about concentration of economic power: In theory they dislike it; in practice they enjoy its fruits.[18]

Out of a total of 111 argument references, 62 were pro-industry, 18 were opposed to government policies, and 31 were anti-industry. Market-share figures, for example, were used 11 times (primarily in the article dealing solely with divestiture) to illustrate the competitiveness and relatively low concentration of majors in the oil business and of the oil business within the total industrial framework. Ease of entry by other firms and joint ventures or product sharing were also arguments—presented from both a pro and a con standpoint—used to discuss the major, multinational oil companies' relationship to other industry members. As one analysis article was careful to point out,

> If sheer market share is the measure of abuse, it is difficult to make a case against the industry. Huge as it is, Exxon controls only 8.5% of domestic production. The top five majors account for barely 35%. There are thousands of small-time drillers; 20 independents have built refineries since 1950; the penetration of nonbranded marketers is up from 18% in 1965 to 30% today. "The sheer statistics," concedes Thomas Kauper, the Justice Department's antitrust chief, "suggest that this isn't one of our more concentrated industries."[19]

The squeezing argument—that majors had cut off independents' supplies and subsidized their own downstream (nonproduction) operations in order to drive independents out of business—was employed six times (refuted twice, simply presented as a contention four times). And the majors' failure to build up oil exploration operations was defended once. Two major areas of concern were profits (which *Forbes* tried to justify 12 times—more than any other area) and the issue of abuse of political power or coercion by the majors (8 times). On profits, one article laments that "[T]he OPEC countries—probably to draw attention from their own outrageous oil prices—are claiming that the international oil companies are making too much money—who's kidding whom. . . ."[20] Another article (March 1975) featuring the views of three major oil company heads argued that the profit spigot had already been turned off and that oil's return on capital was below the average for other U.S. industries. This is easier said than believed, it says, because

> Big Oil's next hurdle will be to convince a hostile public that times have changed. It will be a tough sell—to say the least. You can be sure that some critics will accuse the industry of hidden profits.

Others will simply ignore the numbers and continue to shout anti-business slogans.[21]

The oil industry was chastized for its secretiveness and unwillingness to disclose information on two occasions, but it was seen as a scapegoat for energy mismanagement and misplaced anger (especially from senators playing to the gallery) in seven instances. Responsibility for high prices was placed squarely on OPEC's shoulders 11 times whereas oil companies were blamed for high prices and shortages only twice.

Arguments critical of government policies were used 18 times: the depletion allowance 3 times, intangible drilling expensing once, and general government policies of regulation and/or punitive actions (including threat of divestiture) 10 times. Firm integration was favored as a means of improving firm efficiency and competitive ability 6 times. Whatever arguments were used, their thrust was usually to point to industry's woes in a harshly adverse business climate.

The sentiment for breaking up the oil leviathans is not restricted to the left. Some conservatives are as concerned about giantism in industry as about giantism in government and labor. Big Oil has few outside defenders these days—even though, somewhat belatedly, it is trying hard to polish its image.

Only a small part of the electorate is interested in political philosophy. But as men like Senators Bayh and Abourezk well know, the voters want a scapegoat for inflation in general and high oil prices in particular. The logical target, the Organization of Petroleum Exporting Countries, is beyond the reach of Congress. But what is more handy than International Big Oil, five of whose seven major companies are American? And that steady stream of revelations about political payoffs—isn't it confirmation to the man in the street that Big Oil has abused its size and power?[22]

Another article contends that "It [big business] is becoming the nation's favorite political football. . . . There is a real danger that next year lawmakers may go on the biggest antibusiness binge since 1933. . . . It is easy enough to prove with facts and figures that public opinion is all wet. . . ."[23] Yet the article admits that presenting the facts and figures has not been enough to mollify congressional or public ire. Nevertheless, criticism is preferable to regulation: "Admittedly, the prospect of regulation is not a happy one. It was federal regulation of natural gas that helped produce the current shortages. . . . Still, the public is demanding action and is resentful of the past favors showered on the oil industry."[24] And even regulation, including divestiture, is far more acceptable to what the *Forbes* writer perceived as the greatest threat—a government oil or energy company. A main problem, according to the one editorial on press relations and public attitudes, is the communications abyss. "The public doesn't understand the oil industry and the oil industry doesn't understand the public. Possible disaster stares us in the face and nobody is getting through to anyone else."[25]

In presenting predominantly favorable views toward industry's problem of making the facts not only known but believed, *Forbes* enlisted the aid of an interesting assortment of sources. Twenty-one of these represented the oil industry (15 from major oil companies, 3 from middle-sized independents, and 2 industry consultants), 20 were government sources (13 senators, 3 representatives, and 4 government agency officials), 14 were academics or independent consultants (including Walter Levy, Paul Frankel, and M. A. Adelman), 3 were critical of industry in general, and 4 were news sources—in this instance, also viewed as industry critics. The most frequently cited congressional sources were two Republicans and one Democrat opposed to too hasty regulation of industry. Henry David Thoreau, Ralph Nader, John Kenneth Galbraith, Robert Heilbroner, Leonard Woodcock, and four anonymous news sources were presented to illustrate the various criticisms to which industry has been subjected.

The direction of sources in *Forbes* was quite balanced. Twenty-seven sources were favorable toward the oil industry and five promoted the aims of business in general—a total of 32. Twenty-seven sources opposed the oil industry's status and activities and five expressed a general sort of antibusiness sentiment—a total of 32. The six additional sources were neutral in their biases.

The amount of coverage granted to the different types of sources was also fairly well distributed on the direction continuum. Three articles gave predominant coverage to pro-industry sources, two highlighted the views of anti-industry sources, and three balanced source coverage in terms of physical space.

Although there was careful attention to a balance of sources, there is no question as to *Forbes'* sentiments. Business in general has gotten a "bum rap" (three articles) and big business has been under particularly heavy and unwarranted fire (three articles). There was little indication of any strong attitude in either direction toward small businesses or independent oil companies, but big oil was treated very positively in three articles and positively in two others. Th The market mechanism was highly favored in five articles as a means of setting fair prices and providing industry with the incentive it needs to continue to foster economic growth. Conversely, government interference in this process was viewed as anathema in five articles. The prospect of a government oil company was anticipated with extreme alarm in one article, as were price controls; but there was surprisingly little to-do about an excess profits tax (one article applauded the decline in tax breaks for the oil industry. The oil industry as a whole was treated very favorably in five articles and favorably in a sixth. Five articles clearly expressed the view that the industry is competitive (though except for one article on divestiture, little actual source material or data were used to back this claim), and three articles voiced opposition to divestiture as a means of restoring competition (with vertical integration believed to provide actual economies of scale).

In sum, most of the attitudes and opinions expressed in *Forbes* were very strongly and plainly articulated in its so-called analysis articles. There are few doubts as to where *Forbes* stands and who its most appreciative readers must be. Nevertheless, this is a spirited publication, employing low-profile humor and sarcasm more often than hard facts. However, *Forbes* did pay considerable attention to economic experts' views and focused more on traditional economic indicators of interfirm competition than did the previous two business publications. Its choice of additional sources to prove certain points was ingenious, catholic in taste, and definitely lively reading. But the balance that was built into source choice was largely destroyed by argument selection and by this publication's overriding conviviality with business. Nonetheless, there were serious undertones urging business to get back in touch with the citizenry—a wistful desire for public harmony and free enterprise (recognized as improbable) all rolled up in one.

FORTUNE

This monthly business periodical published five articles on oil industry structure and behavior and related topics in April 1973 and March, April, and December of 1974. Two of the articles, an analysis and a book review, were by *Fortune* staff members; the remaining three were unsigned editorials. Only one of these articles, the analysis, discussed divestiture legislation per se. Other topics covered profits (in two articles), the need for candor in meeting criticism of the oil industry (in two articles), the oil companies' need to improve their investment capabilities through profits and without fear of government reprisal (in three articles), what the U.S. government should be doing about OPEC (in one article) and the role of OPEC in shaping the world petroleum market (in one article).

Fortune utilized 44 arguments to demonstrate its interpretation of how the oil industry, especially major and multinational companies, has conducted itself. Largely because of the lengthy analysis (10 of a total of 13¾ pages for all five articles) and in contrast to what might be expected in a case where three out of five articles are editorials, a number of anti-industry as well as pro-industry views and arguments appeared. Nontraditional conduct arguments tended to receive greater attention than market share, ease of entry, or more traditional conduct measures.

Twenty-three arguments were offered as criticisms of the oil industry. Among the criticisms that were presented and not refuted by contrary data were squeezing (once), collusion among majors (twice), cooperation with government in creating an anticompetitive market state (five times), oil companies giving in too easily to OPEC (once), reporting and accounting gains (twice), industry's poor image (three times), the oligopolistic nature of the majors (once), secretiveness (twice), and having too much power and privilege (once). The book review, for example, quoted economist M. A. Adelman as saying that the OPEC price-fixing cartel was "the greatest monopoly in history" (an argument that offers some defense against the notion that the oil companies are monopolists). But it also quotes Adelman's contention that the oil industry has, in part, allowed an OPEC takeover and has received some advantages from this event (an anti-industry argument): ". . . where marketing conditions allow companies to pass along cost increases plus markup, higher costs need not bring any reduction in profits."[26] However, a similar argument invoked twice—that the oil companies had either caused or failed to avert avoidable price increases—is treated from both a pro and a con standpoint.

Twelve arguments supported the oil industry's position directly and nine opposed government policies that were thought to be either unfair to the industry or no longer purposeful, given current market conditions. The most prominent of these related to profits as both fair and necessary. As one article contends,

> Whatever figures one uses, it is surely clear that the industry will need a large supply of capital and will require handsome profits to get it. . . . As long as that scenario [that supply can be raised rapidly and substantially] seems a possibility, the arguments for allowing price to work its incentive are very persuasive.[27]

Another article, an editorial, stresses the need for candor and reform but also seeks to counter price controls and other measures being sought by certain politicians (the article calls them demagogues).

> Events have transformed the oil companies into a special kind of private enterprise—one that performs a function so crucial to society as to require close public scrutiny and a measure of public supervision. The critical question is how much and what kind of supervision. . . . Having invited the demagogy by neglecting to keep the electorate informed about their business, the oil companies must now make an extraordinary effort to educate Americans about the economics of petroleum. Fundamentally the industry has to persuade the nation that only by remaining highly profitable will the companies have the financial muscle to find more oil and develop alternative sources of energy.[28]

In sum, profits are defended eight times and a windfall profits tax is criticized openly once. The remaining arguments support the notion that OPEC is the major cause of high prices and shortages (four times) and berate the government (nine times) for being ill informed, causing unnecessary delays in easing shortages, failing to enforce certain laws (such as unitization) or to eliminate others (such as depletion allowances, foreign tax credits), or failing to redesign unwieldy and ineffective mechanisms (for example, price controls, allocation formulas). To summarize, *Fortune* incorporated 12 directly pro-industry and 9 antigovernment arguments (the latter tending to support industry's position) for a total of 21. This is almost in perfect balance with the 23 anti-industry views presented.

The sources *Fortune* calls upon are also nearly evenly divided among industry sources (cited five times), academics or independent consultants (cited seven times), government sources (cited or mentioned seven times) and news sources (cited six times). *Fortune* relied largely on cited sources (25) as opposed to mentioned (4).

All five oil-industry source references were associated with major or multinational oil companies—none with independents. The consultants (two academic economists and one government consultant on petroleum—M. A. Adelman, Walter J. Mead, and Walter J. Levy) were cited for comments that both supported and criticized industry's actions. The government sources included two senators, both annoyed by oil company profits (one was quoted twice for saying, "the oil industry is selling less and making more "),[29] and three government agencies or regulatory bodies. The news sources were four political cartoonists whose anti-industry cartoons illustrated the analysis article, and a defender of the industry's right to go unhampered by government, William F. Buckley. Buckley is quoted as saying, "A society that tolerates demagogy is unlikely to penetrate economics."[30] Ten of the sources represented the oil industry's view, nine were in opposition to the major oil companies' activities and structure, and six remained neutral—again, a fairly well-balanced presentation.

The amount of space granted the different sources, on a per article basis, tipped slightly toward the industry. The book review devoted space primarily to academic sources (although a multinational company head was also quoted to illustrate a discussion that blamed OPEC for U.S. shortage and price woes). Furthermore, these academics' views were split in terms of supporting or opposing an industry position. The analysis article gave more space to sources with anti-industry sentiments. Of the three editorials, one gave more space to pro-industry sources and two featured anti-governmental-regulation sources. But since the analysis article was much lengthier than the editorials, *Fortune*, in fact, gave greater coverage to anti-industry sources (followed by neutral sources) than to those who support the aims of the major oil companies.

A glimpse of *Fortune*'s tone or bias can be seen in the attitudes and opinions category. But to assess these fairly requires a consideration of the kinds of articles *Fortune* ran. Three out of five were editorials and did, indeed, reflect biases. The remainder, however, appear to have been written with much effort toward balance, if not total objectivity.

Given that caveat, *Fortune*'s position on several issues can be summarized as follows. Business in general was treated very favorably twice and neutrally or not at all three times. Virtually no attention was paid to big business, small businesses, or independent oil companies. Major oil companies were treated very favorably in one article (in the sense that it commiserated with their current legislation/regulation plight), only slightly favorably in two articles, and non-committally in the other two. Most of *Fortune*'s voice seems to have been reserved for free market versus government control issues. It is quite clear in four of the five articles that *Fortune* thinks the laws of supply and demand should be allowed to work their way without the following (all greeted very unfavorably): price controls (criticized three times), allocation (once), a windfall profits tax (once), the depletion allowance (twice), the maximum efficient rate—a formula for controlling oil production—(once), foreign tax credits (once) and drilling expensing (once). However, *Fortune* did propose two government measures that it thought would further conservation efforts: unitization laws (which, by physically limiting the number of well operations in a field, preserve oil pressure and thus maintain supply longer and more economically), and a gasoline tax with offsetting credits for the poor (designed to discourage consumption). Three of the articles were quite favorably disposed toward the oil industry in general and four articles indicated confidence in its competitiveness. Only one article voiced moderate concern that the Senate would enact divestiture legislation.

To conclude, *Fortune* magazine covered the divestiture/monopoly issue very sketchily between 1973 and 1975. Its major focuses were, instead, on the need for the oil industry to continue its recent good profits and a defense of these profits; the need for price or other market stimulants to further oil exploration and development; and the need to adopt conservation measures and end discriminatory regulations (both pro-industry and anti-industry) so that the oil industry could adjust itself to a changed market situation.

Fortune's coverage was a well-balanced, largely economic approach to the oil monopoly question (as it relates to profits), pausing only occasionally to wince at public outrage or misunderstanding. In addressing the public image issue, this publication's greatest complaint was that politicians had oversimplified issues and needlessly fanned consumers' contempt. "Politicians," it notes, "are quick to sense possibilities for political profit."[31] Taking a cue from one of its own editorials titled, "Oil and Emotion Don't Mix," *Fortune*'s approach was to counter these "political" attacks with well-documented (albeit somewhat limited) arguments in a calm, analytic manner.

SUMMARY OF RESULTS

The business periodicals analyzed—*Business Week, Dun's Review, Forbes*, and *Fortune*—all examined some aspect of oil industry competitiveness and/or divestiture's implications. These publications also devoted considerable space to discussing the criticisms to which the oil industry has been subjected, the significance of oil industry profits since the embargo, and the effects and advisability of various government policies that affected the industry. In addition to identifying these periodicals' concerns, the content analysis was particularly concerned with identifying the types of sources employed (especially whether inclusive of independent consultants, academic economists, or similarly impartial or objective sources), the kinds of arguments presented, and the extent to which these publications balanced source and argument presentation.

Business Week provided a number of brief, largely news articles. Sources were comprised mainly of industry figures followed by governmental and nongovernmental industry critics, with only a sprinkling of independent, academic, or nonindustry consulting sources. Sources were split among pro- and anti-industry views, tipping slightly toward anti-industry sources. Slightly more source space was devoted to pro-industry sources. However, arguments critical of the oil industry numbered nearly twice as many as those supporting industry's contentions—perhaps a product of a news-oriented focus on FTC and Senate antitrust activities.

Dun's published only three analytic articles relating to oil antitrust (and larger antitrust) matters. These favored pro-industry sources both in numbers and (slightly) in terms of space allotted different sources. However, pro- and anti-industry arguments were fairly well balanced and the proportion of independent sources was higher (about one-fourth of all sources) than in *Business Week* (whose proportion of independent sources was only about one-tenth).

Forbes published eight medium-length (two- to three-page) articles dealing largely with industry criticisms and with profits. It balanced pro- and anti-industry sources quite evenly. It included fairly even proportions of business and government sources and about one-fourth each of independent sources and nongovernmental critics. Source space tended to favor oil industry sources (as did authors' commentary), and arguments favoring the oil industry's views outnumbered anti-industry arguments about 3 to 1.

Fortune provided the most balance in all respects—a surprising result considering that three of its five articles were editorials. There was also a larger proportion of sources (about a third) whose views were neither pro- nor anti-industry but represented an analytic approach to government energy policy. Its small number of total sources were evenly divided among independent consultants, business, government, and nongovernment sources. Again, arguments and sources were almost evenly balanced, with slightly more source space devoted to industry critics.

NOTES

1. "Independent Oilmen Fight for More Freedom," *Business Week*, June 15, 1974, pp. 61, 65.

2. "The Future of Oil Profits," *Business Week*, February 2, 1974, p. 76.

3. Ibid.

4. "Independent Oilmen Fight for More Freedom," op. cit., p. 61.

5. "Mobil's Warner: An Oil Industry View," *Business Week*, December 22, 1973, p. 115.

6. "A Bookkeeping Maze Behind Oil Profits," *Business Week*, May 4, 1974, p. 23.

7. "The New Shape of the U.S. Oil Industry," *Buisness Week*, February 2, 1974, p. 55.

8. Ibid., p. 58.

9. "Oil Companies Get Hit by a Poll," *Business Week*, March 16, 1974, p. 24.

10. "Realism on Prices," *Business Week*, September 28, 1974, p. 116.

11. "An Energy Compromise," *Business Week*, March 31, 1975, p. 80.

12. Lee Smith, "Oil Meets the Press," *Dun's Review*, April 1974, p. 62.

13. Ibid., p. 124.

14. "Close Up: The FTC Curbs Competition," *Dun's Review*, July 1973, p. 13; and Jean Ross-Skinner, "Can We Crack the Oil Cartel?" *Dun's Review*, February 1975, p. 58.

15. Smith, op. cit., p. 122.

16. "Big Business: Hard to Live With—Impossible to Live Without," *Forbes*, May 15, 1974, p. 49.

17. Ibid.

18. "The Oil Giants: An Irresistible Target," *Forbes*, January 15, 1976, p. 22.

19. Ibid.

20. "The Mote and the Beam," *Forbes*, December 1, 1974, p. 18.

21. "Oil's Big Spill," *Forbes*, March 15, 1975, p. 21.

22. "The Oil Giants," op. cit., p. 20.

23. "Business on the Run," *Forbes*, May 15, 1974, p. 54.

24. "The Way the Wind is Blowing," *Forbes*, May 15, 1974, p. 56.

25. "Why Are They Picking on Us?" *Forbes*, April 15, 1974, p. 7.

26. Gilbert Burck, "The Greatest Monopoly in History," *Fortune*, April 1973, p. 154.

27. Carol J. Loomis, "How to Think About Oil Company Profits," *Fortune*, April 1974, p. 198.

28. "Countering the Demagogues," *Fortune*, March 1974, pp. 99-100.

29. "Oil and Emotion Don't Mix," *Fortune*, April 1974, p. 96.

30. "Countering the Demagogues," op. cit., p. 100.

31. "Oil and Emotion Don't Mix," op. cit., p. 96.

NEWSWEEK

This weekly news magazine published 23 articles or a total of 18¾ pages on oil industry competitiveness, and pricing and supply issues from July 1973 through October 1975. Seven articles listed no author. The most frequently cited author was James J. Bishop, Jr. who wrote or collaborated on seven of the articles. *Newsweek* economics columnists Milton Friedman of the University of Chicago and Paul A. Samuelson of M.I.T. wrote four and three viewpoint articles, respectively, on oil economics. The remaining articles were by various editors, correspondents, and an outside contributor. Seven were straight news articles, seven were news/analyses (more in-depth looks at events in the news), seven (Friedman's and Samuelson's) were opinion/analyses, one was an interview, and one an opinion/satire (in the "My Turn" section by outside writers).

Only two of these 23 articles dealt directly with divestiture, one solely and one predominantly. The topic most discussed was the energy crisis or oil shortages, high prices, who's to blame, and what does it portend (ten times), followed by the effects of price controls and allocation programs (six times); industry misconduct or privilege issues, such as conspiring to create shortages, or the tax, profit, and accounting benefits enjoyed by the oil industry—especially majors (five times); other criticisms directed at the oil industry such as withholding information (three times); and Congress's slowness to alleviate energy problems (once).

The arguments incorporated in *Newsweek*'s articles included 57 that supported the oil industry's basic views (21 of which blamed OPEC for this nation's energy problems), 98 that were critical of oil companies (almost exclusively

majors or multinationals), 4 opposed to government inaction, and 6 critical of
various government policies (but not used in either a pro- or anti-industry way).

The anti-industry arguments most frequently incorporated were that major
oil companies controlled supplies, and thus were able to squeeze independent
producers, refiners, and marketers out of competition (11 times); that the major
oil companies had reaped windfall profits or had made extra profits through inventory or accounting mechanisms (13 times); that major, integrated firms had
colluded either with OPEC, the U.S. government or, usually, with one another to
achieve squeezing and other anticompetitive effects (10 times); that industry had
either created shortages or failed to warn the public of their impending occurrence (10 times); that the oil companies or major oil companies were responsible
for high prices (6 times) and shortages (11 times); and that the industry perpetu-
ated or took advantage of government policies that improved its financial position (7 times).

Below are some examples of how these anti-industry arguments were incorporated into *Newsweek* articles. On shortages (who or what caused them and
what have been their effects) and on squeezing, two separate articles say:

> With the gasoline shortage becoming worse each day, growing num-
> bers of Federal and state officials are asking: Is the shortage legiti-
> mate? Or is it due to a conspiracy by oil-industry giants to drive
> competitors out of business, boost their own profits and win major
> concessions . . . evidence of outright collusion is as scarce as an open
> gas station on a Sunday night. But there are indications that many
> major oil companies have at least used the shortage as an excuse to
> cut off independent refiners and retailers from supplies of crude oil
> and gasoline.[1]
>
> . . . These integrated activities have contributed to the short-
> age of oil products, critics say, by forcing many independents out of
> business.[2]

On collusion among industry members the previous article also points out that:

> Few detractors believe that the oil giants meet furtively in lavish
> boardrooms or hotel suites to plan a collective strategy. "They don't
> have to," says Ralph Nader. "They are like Siamese twins. They
> know exactly what they are doing without meeting." What they do,
> according to Nader, is to exacerbate a shortage situation by under-
> stating their reserves and withholding oil from the market in order to
> push prices up and force a retreat from environmental laws. And
> the only estimates of oil reserves the government has are those sup-
> plied by the oil companies themselves.[3]

Big oil's influence-wielding effect on government policies (now deteriorating) is
addressed in another article:

Americans have never felt kindly toward the oil industry, but the oilmen have historically had enough friends in Congress to prevent the
passage of any really painful restrictions on their growth and profits.
With everyone suffering nowadays from higher oil prices, however,
all of the old bitterness has again reached a peak and suddenly the
Senate is engaged in a serious drive to break up the oil industry. . . .
The very fact that the industry even feels threatened by a divestiture
bill is a sign of the changing times and the decline in the oil lobby.
Oil influence in recent years hasn't been as powerful as it was in
1931, when Texas Gov. Ross Sterling, a former president of Humble
Oil and Refining Co., ordered state troops to close new wells in
order to keep oil prices up. But as recently as 1969, Exxon Corp.
was able to use its influence with President Richard Nixon to prevent the building of a refinery at Machiasport Maine, which would
have produced cheap oil for New England.[4]

Another article, a satire/opinion piece from an outside contributor, suggests that
the majors have cooperated with the oil exporting nations. [Says "Offshore
Harry"], "It's not conspiracy exactly, it's just that this generation of Arabs has
learned it can push the companies around, and the companies don't have to push
back very hard, do they?"[5]

On profits, there were a number of comments and arguments. One article,
for example, written in late 1973, stated that the price controls argument of
industry is "astounding in light of oil profits this year."[6] Another reported in
early 1974:

It seemed truly an embarrassment of riches. Just as U.S. oilmen were
telling Congress last week that they were innocent of a series of ac-
cusations—including profiteering from the energy crisis—the companies began reporting results for 1973 that seemed to bolster the case
against them.[7]

However, the report did add, "The danger is that in its anger over high prices and
profits, Congress might over-react." On the subject of "funny" profits or reve-
nue accrued because of accounting or pricing procedures, one news article notes:

According to the FEA, some large international companies have
been overcharging themselves when they pass crude from a foreign
subsidiary to a U.S. affiliate. This "transfer pricing" inflates the
profits of the foreign subsidiary while the overcharges are passed on
to U.S. customers. . . . The overbilling takes place, FEA says, because the companies use the accounting techniques most advantage-
ous to them. . . .[8]

Newsweek often highlighted the many governmental investigations seeking answers from (or about) the oil industry. In describing these inquiries, the

publication examined the conspiracy theory—which argues that the industry trumpted up shortages or cooperated among industry members unfairly—and offered some skeptical comments' on Congress's eager delving. One article said in 1973:

> . . . it quickly developed that this wasn't going to be just a dry recitation of petroleum statistics. Instead, for the next three days, the American public was treated to an often acrimonious, finger-wagging, paper waving spectacle in which the giant oil firms were under the gun as never before. . . . As the shortage dragged on, Congressmen were launching so many investigations that matters seemed to be getting out of hand. Indeed, the Senate Democratic Conference named a committee to determine whether one committee should be named to conduct the proliferating number of probes.[9]

And almost a year later, another article admitted:

> Whatever the investigations eventually turn up, there has been little if any evidence so far that the current shortages are trumped up. . . .
> In the end, consumers may have to quit fretting about the possibilities of conspiracies and get used to the idea of paying unheard-of prices to run their cars and heat their homes—if they can get fuel at all.[10]

There were some arguments that tended to support industry's contention that it was innocent of collusion and had behaved competitively. The primary one was that OPEC is the body that bears the main responsibility for higher prices and production cutbacks (21 times). This was followed by arguments that government policies—especially price controls, allocation formulas, and quotas—have worsened the effects of the embargo-produced price rises and tight supplies (15 times); that the industry must have profits if it is to make the investments needed to expand new, secure supplies (5 times); that the oil industry is not unduly concentrated compared to other U.S. industries and does act competitively (2 times); that the industry did try to warn the public that shortages were imminent (2 times); and that oil profits did achieve a temporary rise but, on balance, have not been excessive (2 times).

The most frequently used pro-industry argument is actually an anti-OPEC argument. A *Newsweek* economic correspondent says of OPEC, "[A] monopoly that is not content with mere gouging but insists on bringing its victims to their knees drives them to increasingly radical counteraction."[11] The OPEC monopoly is treated as a separate, foreign-policy issue, rather than as related to the domestic oil situation. Several articles, however, do discuss the effects government policies, particularly of price controls and allocation, have had on domestic supply. According to the previously cited economic correspondent,

The energy managers . . . were working their way around the Congressionally mandated allocation system for crude oil, which admittedly has worsened shortages. For one thing, it forces crude-rich companies to share their oil with competitors, thus discouraging the latter from importing additional supplies.[12]

All of Milton Friedman's columns support at least one aim of the oil industry—a removal of price controls or other mechanisms that interfere with the market's functioning. He blames OPEC for high prices and argues in what he calls "A Dissenting View" that:

Decontrol will *not* produce a sharp immediate rise in the price of petroleum products, it will reduce the power of the OPEC cartel to impose higher prices on the U.S., and ultimately it will lower prices as the free market works its magic. Decontrol *will* produce drastic redistribution of profits within the domestic oil industry. Price control imposed a heavy tax on some domestic firms, granted a large subsidy to other domestic firms and subsidized the import of oil.[13]

Economist Paul Samuelson, too, although much less willing than Friedman to allow unrestrained market forces, agrees that OPEC is the chief monopolist causing shortages and high prices, and favors gradual price decontrols.

Where lies the golden mean of policy?. . . First, a slow phasing out of controls offers a good, perhaps the best plan.

Second, even instantaneous decontrol needn't negate healthy recovery. The Wharton forecasting model ran a September calculation on "instantaneous decontrol." It lowers the real growth forecast from mid-1975 to mid-1976 by 1 percent; raises Election Day unemployment to only 8.1 from 7.9 percent . . . and raises the price level on Election Day by only 1 percent.

Moral: The energy debate is unnecessarily hot.[14]

Another article places the blame for shortages not only on OPEC but on congressional "do-nothingism" on energy. The article charges that rather than take effective action, members of Congress chose to listen "to what they perceived to be the voice of the people back home." As a result,

They killed proposals to increase the gasoline excise tax, softened provisions for oil-import quotas and for penalizing gas-guzzling cars. The watered-down version of the bill . . . represented a bitter defeat for the [House Democratic] leadership, particularly Chairman Al Ulman of the Ways and Means Committee, the chief proponent of tough legislation.[15]

The article also quotes Exxon's senior vice president, William Slick, as saying, "They [the legislators on Capitol Hill] still believe in a free lunch up there."

Industry's concern about bearing the burden of blame appears in several articles. One quotes an oil official (also cited in another article in which he claimed that "we *did* warn the public of shortages")[16] who is dismayed by governmental stirrings against the industry:

> "It is tragic, really," says Frank Ikard, president of the American Petroleum Institute. "The mood seems to be for punishing somebody for this. It will make the problem worse." But if the shortage lasts the three years that many experts predict, some kind of legislation against the industry—perhaps a lower depletion allowance, or even a specific antitrust law mandating breakup—seems all but inevitable.[17]

A State Department official also is cited for his opposition to a "witchhunt," arguing that "everybody's to blame."[18]

On the whole, the anti-industry arguments were more pointed and plentiful than those supporting industry's views. As one article put it, "Predictably, the oil companies struck back at the charges of conspiracy."[19] But often these counterarguments, such as an oil executive's single comment, "Nonsense," were presented rather perfunctorily. *Newsweek* focused heavily on the conspiracy theory for quite some time before concluding that there was not much proof behind the charge. As a result the reader received a lot of exposure to the anti-industry side of the issue. However, there was a gradual shift in *Newsweek*'s presentation. Early articles focused a great deal on shortages and blame laying. Later articles tended to be more analytical and more questioning of some of the causes of the energy crisis (later revised to read "problem"), including OPEC's role, people's patterns of consumption, and government policies, as well as industry's contribution to the problem.[20]

In sum, then, a substantial portion of the arguments in *Newsweek* were critical of industry's role in creating and perpetuating an energy crisis. Later articles were still weighted toward anti-industry arguments but there was more attention to government's failure to take decisive steps either to curb the companies' power or to enact legislation that would speed domestic oil development and, to some extent, reduce consumption. A few articles focused on OPEC's role in worsening the crisis situation. In 1975, when shortages had eased somewhat, *Newsweek* refocused its attention on attempts to strip industry of some of its previously legislated benefits or to pursue antitrust actions.

Newsweek cited 88 sources and mentioned 30 for a total of 118. The most frequently quoted or mentioned were government sources—20 officials of various agencies such as the Federal Energy Administration, the Federal Trade Commission, and the Securities and Exchange Commission; 30 congressional sources;

and a scattering of White House and state-level officials for a total of 60. All 27 industry sources were cited directly. Of these, there were 12 major, 5 independent, 1 middle-sized integrated, and 7 industry association sources plus 2 industry consultants. There were ten references to academic or independent consultants, three to public opinion polls, ten to industry critics, and three to incidental sources. The most frequently quoted sources were Frank Ikard, the president of the American Petroleum Institute (five times); Walter J. Levy, a world oil consultant (an independent as opposed to industry consultant—cited five times); unnamed industry critics (four times); William Simon, head of the Federal Energy Office (three times); and a senior oil official (three times). Senator Henry Jackson was most frequently mentioned (four times) but cited only once. The remaining sources appeared only one or two times. *Newsweek* had a fairly large number of unnamed sources (17) referred to variously as experts, critics, Senate and oil Spokesmen and the like.

Source direction, like arguments, also fell heavily in favor of the critics of the oil industry, particularly the majors. Thirty-seven sources made comments supporting industry motivations or actions. Conversely, 36 were critical of industry and 26 were specifically antimajor—a total of 62. Nineteen source references maintained a neutral stance vis-a-vis the industry, but nine of these favored government action in the form of negotiation with OPEC.

The amount of space given to different sources reveals a strong pattern of balance in 10 of the 23 articles. The remaining articles favored only one type of source, but cumulatively, these, too were fairly well balanced. Two articles gave more space to anti-oil-industry sources; three focused on sources whose statements or actions were specifically antimajor; two gave greater coverage to pro-industry (predominantly major) sources; five favored decontrol measures that industry also favors; and three were critical of OPEC (a neutral position toward the oil industry). However, if deliberate opinion articles (the economic column and the satire) are eliminated from this list, the news and news/analysis articles favored anti-industry over pro-industry sources 2 to 1.

Newsweek, as one might expect of a news magazine, maintained a reasonable, but certainly not a pristine neutrality in presenting actual judgments or commentary. Big oil was treated quite unfavorably in only one article—the satire. It was treated unfavorably in two, mildly unfavorably in five, and neutrally in 15 articles. Independents were treated neutrally for the most part, although two articles championed allocation as a means to help otherwise-strapped independents and two others were somewhat favorably disposed toward independents' perspectives. The remainder were either neutral (11) or did not discuss independents (5).

The market mechanism was viewed highly favorably by Milton Friedman five times and by Paul Samuelson once, favorably in two articles, and a bit less favorably in three late 1974 articles leaning toward modified or partial price decontrols. The market mechanism was viewed very unfavorably in one, unfavorably in two, and neutrally in ten articles.

Government action to alleviate shortages and ease the price burden was viewed favorably four times and partially favorably twice. Government action in general produced three very unfavorable responses and one mildly unfavorable response from *Newsweek*'s resident economists, Friedman and Samuelson, respectively. In this same vein, price controls received six very bad marks plus one unfavorable and one favorable; the rest were neutral or, more often, not applicable. A "Do-Nothing Congress" (or Administration) banner was waved in four articles—one very unfavorable, two unfavorable, one slightly unfavorable. Government intervention to negotiate with OPEC, however, was posited as a good strategy in two articles (one quite favorable, the other only mildly favoring such a step). But it was discarded as not too helpful in two articles and not treated at all in the rest.

As for other possible government policies, taxes on energy use were treated favorably twice; allocation was highly favored once, highly opposed twice; tax benefits for industry were viewed very unfavorably three times; and investigations into oil companies' monopoly power were favored once. (It should be noted that *Newsweek* ran a great many articles, not covered here, on investigations into other types of oil company abuses, particularly overseas bribery charges.)

The oil industry as a whole was handled neutrally 16 times, mildly to somewhat more favorably 5 times, and mildly to somewhat more unfavorably twice. The question of the competitiveness of the industry was raised in nine articles with two concluding that it definitely was not competitive, three leaning mildly in that direction, two considering it competitive, and two considering it mildly competitive.

Divestiture was treated neither as a solution nor a disaster. Slight mention was made of divestiture legislation in only two articles and treatment was neutral.

To summarize, the way in which arguments and sources were presented in *Newsweek* tended to show that the forces of the oil industry and, to a lesser extent, OPEC had worked to create shortages and high prices. However, OPEC's supply and pricing were seldom linked to those of the domestic oil companies. Economists Milton Friedman and Paul Samuelson, in fact, were the only authors who seemed intent on illustrating this relationship. Perhaps the compartmentalized presentation one senses in *Newsweek* is inevitable in a magazine, which, of necessity, looks to newsmakers to provide brief story themes. Generally, those newsmakers have one fairly narrow view that they present to the exclusion of others. The difficulty of capsulizing source views is probably another deterrent to looking at a story's several facets.

At any rate, in terms of arguments and sources, *Newsweek* achieved better balance among articles than within them. The problem with this approach is that different readers might select or ignore articles developing certain themes. *News-week* has tended then, to segregate or group articles into certain categories—major oil companies, Congress's role in energy, OPEC pricing, and the problems of independents (in that order). Cross-breeding occurs not so much by design as when events, such as congressional investigations or hearings involving majors' testimony, actually bring the principals together. Although *Newsweek*'s tone is relatively neutral, its bias toward anti-industry arguments provided by anti-industry sources is plain. The sources do tend to be mediagenic. Perhaps that is why they are chosen and why their views seem to shine through. *Newsweek*'s approach makes for interesting but not always balanced or analytical reading. Perspectives within articles are limited, and many articles emphasize a growing distrust of the major, integrated oil companies, the oil exporting countries, and the U.S. government without presenting the views of relatively disinterested or dispassionate sources.

TIME

From April 1973 through November 1975, this weekly news magazine published 11 articles or a total of nearly nine pages on divestiture, oil industry practices, and related topics. None of the articles received a by-line. Six articles, focusing mainly on shortages and alleged oil company practices, were straight news; four were news/analyses—two on antitrust actions, one on profit accounting, and one on the government's role in formulating energy policy; and one was a personality profile of consumerist and oil industry critic, Lee White.

Time incorporated 76 anti-industry arguments compared to 21 of a pro-industry nature. Its presentations of government policies were used approximately twice as often to illustrate anti-industry as pro-industry positions. Arguments that placed blame for high prices, shortages, or an energy crisis were directed primarily at government policies, followed by the major oil companies, and last, by oil companies in general.

The most frequently appearing anti-industry argument focused on excess profits (ten times), followed by cooperation among industry members to maintain high prices, withhold supplies, or share supplies or products among themselves (nine times); industry's (especially majors') failure to disclose information about its market or practices (six times); squeezing independents (five times); price gouging in general (five times); using devious accounting or reporting devices (four times); noncompetitive market share (twice); and acting as a monopoly (once).

Oil companies or major oil companies were viewed as being responsible for high prices or shortages seven times, whereas OPEC was blamed four times. Government policies thought to contribute to shortages and high prices or to create or allow unfair oil company advantages were tax loopholes (twice), the depletion allowance (3 times), foreign tax credits (twice), allocation formulas (twice), import quotas (3 times), vertical integration (3 times), and horizontal integration (once).

Arguments used to demonstrate pro-oil-industry positions included market share (twice); cooperation within industry as a means of facilitating commerce (once); profits only temporarily high as a result of higher prices, inventories, and so forth (three times); OPEC as the primary cause of high prices and shortages (four times); poor or outmoded government policies that intervene in the oil marketplace (eight times); vertical integration as a means of creating economies of scale and meeting demand (three times); and the occurrence of a shortage situation due to tight world supplies (once).

Although it is possible to separate *Time*'s presentation of pro- and anti-oil-industry statements, it appears more appropriate to present them in the point-counterpoint fashion often employed by this publication. For example, in early 1973 an article on the effects of shortages on independents said:

> Independent marketers, who have captured 22% of the retail gasoline trade, suspect the major oil companies have contrived the shortage to force them out of business, drive up prices, and silence environmental critics. . . . Spokesmen for the major oil companies claim that refining runs are down because their stocks of unrefined crude oil are dwindling in the face of a world wide tightness of supply.[21]

Another, a news/analysis of profits, notes that "Some overcharges were the result of honest misinterpretation of federal pricing guidelines, Sawhill said, but others 'were out-and-out examples of price gouging.' "[22] An article on divestiture efforts states:

> With so much under the companies' umbrellas, the fear that they will exercise a stranglehold over all energy development is not senseless. But it has yet to be demonstrated that they have in fact committed any major sins. In a decade of hearings, Phillip Hart's subcommittee has never proved that the industry practices collusive pricing. [And] . . . critics doubt that breaking up the majors would lead to lower prices.[23]

Finally, another article, pondering whether shortages had been contrived, recognizes that:

> For some time, the nation will probably remain split into two equally hopeful and equally helpless factions, the believers [those who think the shortage is real] and the doubters. And as every American has come to know, a house divided against itself will be a little colder this year.[24]

However, other articles saw little duality in the issues. One news/analysis, for example, favors antitrust actions (whether correct or incorrect) as a way to dispel doubts. " . . . the trustbusting fever will have a beneficial effect: the investigations are likely to determine whether the charges of collusion and contrived shortages are valid—or just a lot of gas.[25] Another also acknowledges the appropriateness of governmental inquiry.

> The huge corporations that dominate the U.S. oil industry have long occupied one of the most lucrative areas in the American economy. Because they are so powerful and pervasive, they have been under almost constant Government scrutiny—and the investigations have intensified as a result of the nation's first peacetime gasoline shortage.[26]

And yet another sees profits as a lever against industry's protests that it lacked economic incentives.

> These [industry's post-embargo] earnings are likely to recede only slightly in the future, but many oil analysts agree that rising fuel prices will continue to pump fat profits into petroleum firms. This could be good news for consumers. With their coffers bulging, the oil giants can hardly plead poverty as a excuse for not increasing exploration for new sources that would be immune to Arab blackmail.[27]

Nevertheless, *Time* takes into account some of the changes that affected the oil industry.

> For about 30 years, the U.S. oil industry had all the friends it needed in Washington and little trouble winning favorable policies: maintenance until early this year of the lucrative oil-depletion allowance, now discarded quotas on oil imports, tax preferences for foreign and domestic drilling operations. Now a wave of hostility unmatched since the breakup of Standard Oil in 1911 has plunged the oil industry into big political trouble.[28]

It also includes a questioning of the effects of divestiture: "The bust-the-big drive came as the gasoline crisis that fueled it seemed to be abating."[29] Furthermore, there is a recognition of political motivations influencing the move for divestiture: "The companies can expect continued hostility from Democrats who sense quite correctly that they have all the support they need from voters eager to strike back at someone or something for rising prices."[30]

Pro-industry arguments, mentioned appreciably less often than anti-industry ones, tended to be derived from interpretations of what was acceptable behavior for industry (for example, industry profit returns were considered reasonable by an otherwise critical government source); were made in defense of charges that had been brought; or indicated that OPEC or government policies were the indirect culprits or causes of the U.S. energy malaise. As a result, these comments often came as disclaimers at the end of a litany of charges by anti-industry (or antimajor) sources and tended to be brief. For example, a usually talkative head of a major oil firm said of the FTC's antitrust complaint charges, "Baloney."[31] And other official spokesmen also seemed to appear frequently near the ends of articles to deny whatever had gone before.

One article suggests that it is not the companies but members of Congress who have acted irresponsibly. Pointing to "Congress's failure to legislate any tough energy program," and criticizing a particular vote to override a presidential veto of a strip-mining regulation bill, the article complains:

> . . . the vote merely highlighted the inability of the White House and Capitol Hill to come up with such a [coordinated national] policy, or of the Democratic-controlled Congress to draft any sustainable energy program of its own. So long as the deadlock continues, the U.S. will apparently be left to OPEC's none-too-tender mercies.[32]

Oilmen were quoted in several other articles as blaming OPEC for creating high prices, and to a lesser extent, shortages (which were not regarded for long as being of crisis proportions).

In sum, *Time*'s arguments, although numerically weighted against the oil industry in a seven-to-two ratio, retain something of a stylistic balance due to efforts to present opposing views simultaneously. However, this argument simultaneity, while contributing to understanding in some cases, can also stress certain viewpoints to the exclusion of others. Most often, *Time* undermined pro-industry arguments, making them both briefer and less explanatory.

Time used a variety of sources in covering divestiture/monopoly and related topics, using most sources only once or twice each. The most quoted sources were the FTC staff report that stated the case for divestiture of eight major, integrated oil companies (five times); Lee C. White, an anti-oil lobbyist (five times); Frank Ikard, president of the American Petroleum Institute (three times); John Sawhill, deputy to the chief of FEO (three times); and Senator Henry Jackson (twice). The most mentioned sources were unnamed oil executives (five times) and the Standard Oil Decision of 1911 findings (three times). Of these, 20 sources were pro-industry, 3 against government interference, and 3 specifically promajor, for a total of 26. More than twice as many sources had views of an anti-industry (33), antimajor (22) or pro-government-intervention nature—56 anti-oil views in all. In addition, seven sources were neutral. Source coverage, in terms of space, was also skewed in an anti-oil direction. While four articles appeared to grant equal space to proponents and opponents, one article featured mostly anti-industry sources and six articles gave greater space to antimajor sources, Thus, source coverage, on both numerical and spatial counts, favored anti-industry or antimajor sources heavily, with little inclusion of independent analysts (only 2), and heavy reliance on government sources (50), of which 24 were congressional figures mostly favoring anti-industry legislation.

In terms of attitudes or opinions expressed in the articles, *Time* was neutral toward big oil in five articles, mildly favorable in one, unfavorable in one, and slightly unfavorable in four articles. Independents were treated neutrally in eight articles, not at all in one, and mildly favorably in two.

The market mechanism concept was balanced, treated mildly favorably in one and mildly unfavorably in another article, with the rest neutral. Government action in general was not often discussed, but specific actions received some attention. Investigating to get at the truth about the oil industry was advocated twice; rationing was regarded very favorably in one article and mildly favorably in another; price controls were viewed mildly favorably in two articles, as was a federal energy company (once); an excess profits tax was mentioned but treated neutrally; and a gasoline tax, an auto efficiency tax, quotas to reduce imports, oil tariff elimination, and gradual price decontrol were all treated very favorably in only one article—a news/analysis of federal energy policy.

The general tone with regard to the oil industry was very unfavorable in two articles, mildly unfavorable in four, mildly favorable in one and neutral in the remaining four. The competitiveness of the oil industry was challenged in five articles (with one very unfavorable and four only mildly unfavorable), mildly accepted or favored in one article, and treated neutrally in the rest. Only one article hinted (mildly) that divestiture did not appear to be a good solution to energy problems; the rest were either neutral on the subject (4 times) or did not approach it (6 times).

In summary, *Time* skillfully combined facts, innuendo, and selected sources to present the oil monopoly issue. In glancing over *Time*'s own words one gets the impression that every effort has been made to collect opposing views and sources. Upon closer inspection, however, the tapestry is woven with skeins of anti-industry arguments and sources interspersed with threads of pro-industry rejoinders. The authors appear to have been granted leeway to comment upon the news they report, to such an extent that some biases show through. Nevertheless, *Time* has done a respectable job of assembling popular arguments and people and seems to have enjoyed opening the fracas up to public inspection.

U.S. NEWS & WORLD REPORT

This weekly news magazine featured 10 articles or a total of approximately 21¼ pages on divestiture, monopoly allegations, industry's role in creating shortages or meeting demands, and related topics from March 1973 to February

1976. One of these articles was an editorial by the *U.S. News* editor-in-chief; the remaining unsigned articles were two analyses, three interviews, and four news/analyses. One article dealt   solely with divestiture, four partially with divestiture or monopoly charges, three with shortages, two with industry criticisms, and several others partially with U.S. energy needs and constraints upon industry to meet petroleum demands.

The arguments used in *U.S. News* came out slightly in favor of the oil industry with 85 pro- and 79 anti-industry positions. The arguments favoring industry's contentions employed most often were the lack of government incentives (for example, price decontrols) to spur investment and expansion of domestic oil supplies (18 times); OPEC's role in creating shortages and high prices (15 times, whereas oil companies were blamed 8 times); the rise of demand at a time when domestic supplies are falling (6 times); and the good intentions of environmentalists causing a slowdown in exploration and development of new domestic supplies (5 times).

A number of government policies also formed subjects for criticism but they were both pro- and anti-oil-industry arguments. For example, the depletion allowance and foreign tax credits were regarded as no longer beneficial to anyone, a needless subsidy to industry (three times), whereas price controls were regarded as detrimental to oil development four times and a public good only once. Government inaction in formulating energy policies was criticized six times but from a neutral (pro-oil-industry, that is) perspective.

The remaining pro-industry arguments were well counterbalanced, perhaps even overly compensated for, by anti-industry arguments. For example, on whether shortages had been contrived, there were eight pro-oil and nine anti-oil arguments. Likewise, on the issue of expected versus windfall profits, there were 13 pro- and 16 anti-oil-industry (particularly antimajor) arguments. Market share data showed large oil companies to be not overly concentrated three times and guilty of having too great a share three times. Slightly more anti-oil-industry imbalance occurred in that the oil industry was argued to be competitive four times as against ten when it was not. Squeezing was said to have taken place once but also denied once.

The anti-industry arguments that were presented either without or with few counterbalancing arguments were withholding of supplies from the market by majors (four times); cooperation among majors (twice); deceptive advertising by oil companies (once); price gouging (five times anti-oil, once pro-oil); majors driving independents out of business (four times); dividing markets (once); presenting a poor public image (twice); and companies' failure to provide sufficient data about themselves (twice).

If there was one basic theme throughout this series in *U.S. News*, it was that oil companies had been deterred from going about the business of producing and delivering oil and gas—a function that they perform not only well, but competitively, and one that they can perform only if prices are adequate to spur supply. As one article on independent producers notes, "Here in the Southwest 'oil patch' there is no debate over how to solve the nation's energy crisis. A higher price is seen as the key."[33] Another article, an editorial, although recognizing public skepticism of the oil industry as valid, promotes the view that the energy crisis is real and can be relieved only through greater domestic energy development.[34] An analysis that presents industry and anti-industry views fairly evenly asks rhetorically: "Is the 'fuel crisis' genuine or have the major oil companies contrived it in an effort to cut off some wholesale dealers and take over a bigger share of the consumer market? Are higher prices also part of their goal?"[35] Interestingly, in answering these questions, this article sought out sources who represent middle-sized or small companies (or independent allied business concerns), rather than majors, to counter the arguments made by state and federal officials (or congressional members) whose efforts are presumably aimed at undoing big oil's power. Throughout other articles, too, care is shown in labeling monopoly charges "suspected" or calling collusion "unspoken agreements" when presenting anti-industry views.

Another article, a news analysis of the FTC complaint against the eight largest majors, states the details of the case for monopoly charges but counters these heavily with arguments that call divestiture or similar moves into doubt. One article even uses as a source a man who is normally considered an industry foe. In addition to quoting major sources who claim that the industry is highly competitive, it quotes Senator Phillip Hart as saying (in a disgruntled tone) that the "FTC has to prove not just monopoly power, but anticompetitive behavior."[36]

One interview article discusses divestiture proposals directly with both an advocate and an opponent of such legislation. The interviewer's questions appear to be straightforward enough. Though it is probably possible to take interviewees down particular paths, for the most part the temptation to lead into set answers seems to have been resisted. For example, the counterpart to "Why do you advocate that the big oil companies be broken up?" asked of the person favoring divestiture, is "Do you think the big oil companies should be broken up?" Other questions are similarly parallel.[37]

More interesting and revealing than the litany of relatively balanced arguments incorporated in *U.S. News*—presented in nonspectacular and generally noneditorializing style—are the sources it chose to quote extensively to illustrate ideas. *U.S. News*, although it does not exactly have a "plain folks" penchant, pursued a number of infrequently quoted or less publicly recognized sources (a little over a third of total sources): nonmajors, state-level officials, independent consultants, and less frequently limelighted (usually conservative) congressional sources.

Of a total of 102 sources, 98 were cited directly and 4 mentioned. Industry sources totaling 48 were comprised of 25 majors, 11 independents, 3 middle-sized integrated-oil-company (such as Continental Oil) spokesmen, 3 (nonmajor) oil association sources (for example, the Fuel Terminal Operators Association), and 6 industry consultants.

Government sources totaling 44 included 21 federal-level officials or staff (the two most quoted of not only government but of total sources were William E. Simon and his successor at FEO, John Sawhill); 21 senators, including several not often quoted on oil matters in other news magazines, such as Charles Percy (R.-Ill.), Abraham Ribicoff (D.-Conn.), and Carl T. Curtis (R.-Neb.), and two state-level antitrust prosecutors. In addition there were five independent (but no academic) consulting sources, two from the investment-conscious Chase Manhattan Bank, and two OPEC sources.

Source direction was predominantly pro-industry. There were 61 pro-industry sources and 8 neutral or noncommittal in their statements; but since the latter tended to favor investment opportunities, they were essentially pro-industry—a total of 69. Twelve sources were anti-oil-industry and 22 specifically antimajor for a total of 34. Thus, pro-industry sources outnumbered anti-industry sources nearly two to one.

Source coverage in terms of amount of space also revealed a bias toward pro-industry sources. Five articles gave relatively more space to pro-oil sources, two allotted greater space to anti-oil sources, and three were essentially divided on source space. One article, in fact, made a point of equal space by splitting a two-page article vertically down the middle to allow Senator Gary Hart (divestiture advocate) to face off squarely against the chairman of Standard Oil of California, H. J. Haynes.[38]

U.S. News attempted to maintain a neutral posture when it came to providing actual commentary or conclusions, and often succeeded. Big oil was treated mildly favorably in two articles (in the sense that its exploration and development efforts were endorsed), mildly unfavorably in one article (on poor industry image), and neutrally in the rest. Independents were given a fairly high proportion of attention and treated mildly favorably in two articles and somewhat more favorably in a third article.

The market mechanism—that is, prices acting to encourage supply or reduce demand—was highly favored in two articles, favored in two and partially favored in one other, whereas government action in general was treated neutrally. Two specific proposals, an excess profits tax and a federal energy company, were treated favorably and unfavorably, respectively.

The oil industry as a whole was viewed very favorably once, favorably twice, and mildly favorably once. On the question of whether the oil industry is competitive, one article suggested mild doubt, and another was fairly positive that it was. Only one article suggested that divestiture would not accomplish anything and, in fact, might lead to greater price and supply problems. The remaining articles were neutral on divestiture.

In summary, *U.S. News & World Report* has an obvious disdain for government meddling with the conduct of the oil business and approaches this as a problem for the total oil industry, not solely for the major oil companies. In presenting arguments it has brought forth nearly equal numbers of pro and con statements. However, its sources are those whose sympathies lie predominantly either with the oil industry or in opposition to too much government control of economic matters, and these sources are given more than twice as much coverage as anti-industry sources. Nevertheless, nearly a third of the articles do give equal space to opposing views. *U.S. News* will probably not be remembered for its zesty style or choice of colorful sources. But it did present some minoirty or nonmainstream views in an intelligible way; it' gave both sides of the story in a moderately balanced way; and it looked behind the oil economics scene with some ingenuity.

SUMMARY OF RESULTS

The three news magazines analyzed—*Newsweek*, *Time*, and *U.S. News & World Report*—provided a number of brief articles on the energy crisis, or shortages, prices, profits, and suspected causal factors. *Newsweek* gave the greatest amount of coverage to these topics during the period studied. *U.S. News* focused more on the deleterious effects of governmental market constraints than the other two publications. And *Time* gave slightly more emphasis to profits and to people's responses to the energy crisis or problem than the other two magazines.

Newsweek was the most consistent of the three in terms of balancing source direction (about a three-to-two anti- to pro-industry ratio) and source space (fairly evenly divided among source views). About half its sources were governmental, a fourth business and industry, a tenth academic or independent, and the remainder nongovernmental industry critics or miscellaneous. It favored anti-industry arguments nearly two to one, focusing quite a bit on the FTC and Senate charges.

Time favored anti-industry sources and especially arguments with a ratio of more than two to one and definitely allowed greater space to oil industry critics. Its main focuses were on oil profits, charges of collusion, information withholding, and control over the means of production. *Time* featured about the same proportion of government and industry sources as did *Newsweek* but fewer independent consulting sources (only about 2 percent).

U.S. News gave a great deal of attention and space to pro-industry sources (about a two-to-one pro- to anti-oil-industry ratio). However, some of these were independent oil company spokesmen not used as often by the other news magazines. And, unlike the other two publications, *U.S. News* sources were fairly evenly divided between industry and government sources. It featured proportionately more independent academic or independent consulting sources than *Time*, but fewer than *Newsweek*. However, its argument presentation was far more balanced (not quite a one-to-one ratio) than either *Time*'s or *Newsweek*'s.

NOTES

1. "Is the Big Shortage Just Gas?" *Newsweek*, July 2, 1973, p. 59.

2. "Big Oil–Under Pressure," *Newsweek*, December 17, 1973, p. 79.

3. Ibid.

4. David Pauly with James Bishop, Jr., "Big Trouble for Big Oil," *Newsweek*, October 27, 1975, pp. 82-82.

5. 'Adam Smith' (a nom de bourse), "The Great Oil Heist," *Newsweek*, January 28, 1974, p. 11.

6. "Big Oil–Under Pressure," op. cit., p. 80.

7. "The Gusher of Earnings," *Newsweek*, February 4, 1974, p. 65.

8. "Oil: A 'Killing' in Crude," *Newsweek*, February 4, 1974, pp. 67, 69.

9. "Putting the Heat on Big Oil," *Newsweek*, February 4, 1974, pp. 64-65.

10. "Energy: No Shortage of Suspicions," *Newsweek*, January 14, 1974, p. 64.

11. Rich Thomas, "And Why It May Vanish," *Newsweek*, January 21, 1974, p. 39.

12. "Turning a Crisis into a Problem," *Newsweek*, March 11, 1974, p. 66.

13. Milton Friedman, "A Dissenting View," *Newsweek*, August 25, 1975, p. 62.

14. Paul A. Samuelson, "Oil Economics," *Newsweek*, September 29, 1975, p. 74.

15. David Pauly, Henry Hubbard, and James J. Bishop, Jr., "Energy: Do-Nothing Congress," *Newsweek*, June 23, 1975, p. 69.

16. "Who's to Blame for the Crisis?" *Newsweek*, December 3, 1973, p. 89.

17. "Big Oil–Under Pressure," op. cit., p. 80.

18. "Is the Big Shortage Just Gas?", op. cit., p. 60.

19. Ibid.

20. "Turning a Crisis Into a Problem," op. cit., pp. 65-66.

21. Energy: A Federal Oil Firm," *Time*, February 24, 1975, p. 27.

22. "Oil: More Profit and Suspicion," *Time*, May 6, 1974, p. 69.

23. "Oil: Assailing the Giants," *Time*, November 3, 1975, p. 78.

24. "Policy: No Shortage of Skepticism," *Time*, January 28, 1974, p. 31.

25. "Gasoline: Back Come the Trustbusters," *Time*, July 23, 1973, p. 74.

26. "Antitrust: Going After the Oilmen," *Time*, July 30, 1973, p. 55.

27. "Oil: The Pinch at the Pump Begins," *Time*, November 12, 1973, p. 108.

28. "Oil: Assailing the Giants," op. cit., p. 78.

29. "Back Come the Trustbusters," op. cit., p. 74.

30. "Oil: Assailing the Giants," op. cit., p. 80.

31. "Antitrust: Going After the Oilmen," op. cit., p. 55.

32. "Energy: Asleep in the Eye of the Storm," *Time*, June 23, 1975, p. 57.

33. "Oilmen's View–There'll Be Enough Gas and Oil If People Pay the Price," *U.S. News & World Report*, June 4, 1973, p. 27.

34. Howard Flieger, "Change of Oil?" *U.S. News & World Report*, March 26, 1973, p. 96.

35. "The Gas Shortage–How Real Is It?" *U.S. News & World Report*, June 25, 1973, p. 34.

36. "Did the Oil Giants Rig Prices?" (The Charge and the Rebuttal)," *U.S. News & World Report*, July 30, 1973, p. 28.

37. "Both Sides of a Hot Issue: Break Up Big Oil Companies?" *U.S. News & World Report*, February 9, 1976, p. 25.

38. Ibid., pp. 25-26.

6

THE NEWS/OPINION
PUBLICATIONS:
A CONTENT ANALYSIS

THE NATION

This weekly opinion publication published ten articles or a total of approximately 32-1/3 pages on monopoly, divestiture, and related topics from April 1973 through August 1975. Two of the articles were unsigned editorials. The rest—five analyses and three opinion analyses—were by a variety of professional writers and people who had worked in government or economic planning: Robert Sherrill, a Washington-based political writer; Mary Clay Berry, a freelance writer who was a former correspondent for two Alaskan newspapers; Louis B. Schwartz, Benjamin Franklin Professor of Law and Economics at the University of Pennsylvania and the former chief of the decree section, Antitrust Division, Department of Justice; Donald Bartlett and James Steele, investigative reporters for the Philadelphia *Inquirer*; George L. Baker, a freelance, Washington-based writer; Bennett Harrison, a teacher of economic planning in the Department of Urban Studies at M.I.T.; and Representative Les Aspin (D.-Wisc.), who wrote two of *The Nation's* articles on oil matters.

Of 103 arguments presented, 95 opposed some form of major oil company behavior or some form of governmental, OPEC, or media policy that aided or abetted such behavior. The lengthy and varied list of arguments included: squeezing (5 times) or running independents out of business (2); collusion among industry members (4), between industry and government (4), and between industry and OPEC (4) for a total of 12; price fixing or gouging (2); withholding supplies from market or holding up refinery capacity (6); using harassing marketing tactics against independents (1); using joint ventures or mergers to reduce competition (3); acting as a monopoly (6) or oligopoly (1); having too much power and influence (6); feeding public fears about shortages or contriving shortages (3); failure to provide accurate or complete data about oil operations (5); utilizing accounting techniques or fund transfers to other phases of integrated operations (4); and earning excessive profits (6).

Responsibility for high prices and shortages was shared by major or multinational oil companies, government policies, and OPEC. (In addition, one article blamed the media for turning shortages into a "crisis".)[1] Normally, these would not all be anti-industry arguments, but they are so presented in *The Nation* because the large oil companies are purposefully linked either with causing or failing to avert the domestic difficulties wrought by the OAPEC embargo or with arranging with government to obtain exclusive privileges and maintain the anti-competitive status quo. OPEC pricing (coupled with industry insouciance) is blamed 7 times and government policies (lobbied by big oil) 23 times. Those government policies singled out were the U.S. inadequate petroleum reserves policies (2 times), tax loopholes or benefits in general (6), the depletion allowance (2), foreign tax credits (6), prorationing (1), allocation (1), import quotas (4), not enough price controls (1), the inappropriate use of the national security argument to justify quotas (1), and government's too heavy reliance on industry-generated facts and figures (1). Finally, integration is seen as a problem with industry structure that government should take steps to alter. Vertical integration is criticized four times and horizontal integration twice.

The most consistently damning argument appearing in *The Nation* is big oil's insidious (one editorial goes so far as to call it "incestuous") ties with government.[2] Multinationals (MNCs), especially large oil companies operating abroad, are criticized by several authors for seeking and receiving the blessings of government. According to Robert Sherrill,

> U.S. tax laws [such as those allowing foreign tax credits or tax deferrals] are written specifically to give windfalls to the MNCs. It's the kind of credit, masterfully manipulated by the U.S. oil companies operating in the Mideast, that has resulted in infamously low taxes paid by them in this country.[3]
>
> . . . Using the government as a handmaiden in foreign profiteering is, of course, nothing new for U.S. corporations. But one needn't go back to the days when U.S. diplomatic relations with Mexico, for example, were dominated by efforts to help U.S. oil and mining companies. In 1960, when Indonesia began cracking down on a consortium made up of Standard of California, Texaco, Jersey Standard, Socony Mobil and Shell, President Kennedy and the State Department came to their rescue, working out a profit-sharing arrangement with Indonesia that would keep the oil companies in business in that country. American foreign aid underwrote the entire investment of a new British Shell and Jersey Standard refinery in Thailand. Under direct orders from then Under Secretary

of State George Ball, U.S. diplomats pressured Congo into granting a refinery license to Jersey Standard. In recent years, Nixon has practically turned the State Department over to the oil companies to run errands for them in Russia and the Mideast. And so on around the globe.[4]

Another article, an opinion/analysis by Representative Aspin, decries industry's influence over government officials and officeholders.

During the last election, the officers, directors and a few principal stockholders in the nation's oil and gas companies are known to have contributed more than $5 million to the Finance Committee to Re-elect the President. . . . It is therefore not surprising that the Administration's response to the current energy crisis has been basically pro-industry and anti-consumer. While the ordinary citizen is asked to turn down his thermostat, drive slower, cancel weekend auto trips, and generally quit being "Fuelish," the oil companies have not been called upon to make any comparable sacrifices. On the contrary, the energy crisis became an economic and political bonanza for major producers. . . .

In the final analysis, Congress and the American people are left with only one choice if we are to solve the energy crisis: the excessive political power of the oil companies must be brought under control. One comparatively easy and highly effective way to curb big oil's power would be to make sure that political campaigns are financed through public funds—not paid for by private interests like oil men, who ultimately and inevitably seek special favors from the government after their gifts to successful politicians.[5]

An analysis article by investigative reporters Bartlett and Steele states that the multinationals are guilty of either collaborating with OPEC in its quest for higher prices or using the embargo to their own advantage.

The oil crisis is a classic demonstration of multinational corporations in action. The major companies, most of them American, operate beyond the supervision or direction of the world's nations. Even the Arabs don't really know what happens to their oil once it leaves their shores. It is left to the companies to decide where to ship and refine it, what it will cost, which nations will get it and which ones won't. Right now, Americans seem to have come out on the short end of thos decisions.[6]

Aspin, in a 1973 article calling the shortage "big oil's latest gimmick," also argued that the majors held down domestic supplies by running refineries below capacity, thus hurting independent refiners. "If U.S. refineries today matched the summer of 1969's production record the gasoline shortage could be an inconsequential localized phenomenon."[7]

Finally, charges of monopoly power, abuse of power, and too great influence or deception by the oil companies are intertwined in a number of articles. As one analysis of the control of Alaska's pipelines contends, these abuses are continuations of a long tradition of monopoly.

> Pipeline abuses were among the antitrust violations which led in 1911 to the dissolution of the Standard Oil collosus. However, the individual companies which rose from its ashes went right on using control of pipelines to limit completion. In fact, pipeline abuses were among the violations that formed the basis for the federal government's omnibus "Mother Hubbard" antitrust suit against all the integrated oil companies and the API, a suit that was dropped in 1940 so that it would not interfere with the industry's war effort.[8]

Another article, an editorial, fumes about horizontal integration. "Big oil dominates every phase of the energy industry—coal, uranium, shale, as well as oil—and this stranglehold must be terminated as of now."[9] An analysis of governmental thinking on opening up its naval petroleum reserves to oil company bids also complains of the industry's tradition of noncompetition: "Various officials of Interior have said the department would ask for competitive bidding but that doesn't mean much in the oil industry."[10] Another analysis, charging the majors with anticompetitive pricing and holding down supplies, quotes several sources to support this view.

> "The central fact of the petroleum industry," writes Melville Ulmer (*The New Republic*, January 5-12, 1974), "is that the demand for its product is 'inelastic.'. . . So long as the oil barons can maintain discipline in their industry, as they regularly manage to do, it *always* pays to keep prices on an upward track."
>
> . . . Findings such as those I have reported here have even led the conservative Sen. Henry Jackson to conclude that "the fuel shortage is a deliberate, conscious contrivance of the major integrated petroleum companies to destroy the independent refiners and marketers, to capture new markets, to increase gasoline prices and to obtain repeal of environmental legislation." An FTC study leaked to the press in the summer of 1973, but never subsequently published, similarly concluded that "the oil companies have behaved . . . as would a classical monopolist: they have attempted to increase profits by restricting output. This charge received support when the Senate Foreign Relations Committee, in March 1974, disclosed a confidential memo written by economists working for Standard Oil of California back in 1968—long before any Arab boycott threats.

> According to the *Wall Street Journal*, the company economists recommended strong measures to prevent an oversupply of crude oil, including production cutbacks.[11]

An editorial advocating divestiture offers this rationale:

> . . . the major oil companies have too much power; their possession of power is responsible for the apparent malfunctioning of democracy. . . . Nationalization of these companies would not in itself solve the problem. The units of operation need to be broken up; they are much too large and powerful. What is wanted are companies of a size that would make possible close and continuous scrutiny of their operations. At a minimum, we must be able to tax them. Meanwhile, any judgments about "democracy" should be held in abeyance. It is the undemocratic concentration of corporate power that creates the problem, not the momentary failure of people to perceive facts that are obscured by faulty and misleading information.[12]

The Nation used a surprisingly large proportion of pro-industry, pro-oil, and promajor sources to demonstrate points or arguments. Robert Sherrill in particular, in his discussion of multinational corporations, quoted several pro-industry sources. However, although these sources were depicted as presenting their own points of view, these views were simultaneously refuted by authors of *Nation* articles. Thus, although there were 23 pro-industry sources out of a total of 87 sources, these were actually negative pro-industry. In addition, the negative views of some sources who might, in other contexts, support some of the oil companies' contentions, were used selectively to weight arguments in an anti-major direction—a well-tested, and fairly effective, muckraking technique. Below are a few examples of how these negative pro-industry sources were incorporated.

> The oil majors take [government] assistance for granted—to such an extent that sometimes officials like John G. McLean, chief executive officer of Continental Oil Co., publicly complain when the State Department doesn't hop to their bidding fast enough.[13]

The same article quotes several other sources who, it implies, are biased toward or have special-interest attachments to industry or who are just plain wrong. It introduces one source as "generally an apologist for corporate stupidities," and suggests that another source, a Harvard Business School professor, lacks objectivity.

Another article quotes the American Petroleum Institute, "the industry's own lobbyist," as admitting to running refineries at less than full capacity. And although it presents industry's claim, it argues that this claim is unacceptable (and once again, offered up by apologists).

> . . . many industry lobbyists and academic apologists for big oil object that refineries can't possibly run at more than a week or two, conveniently forgetting that during the summer of 1969, refineries worked at more than 91 percent of capacity for ten consecutive weeks.[14]

One article, citing M. A. Adelman's estimates of the costs of delivering Alaskan oil to U.S. markets, uses this information to make the point that the majors are price setters (whereas the source would be unlikely to reach this conclusion—certainly not by using this particular piece of evidence).[15] Another article lumps together domestic oil industry privilege measures (for example, prorationing, quotas) with protectionist (Adelman's term), autarchic influences in other countries. These influences, the article's author argues, led to artificially high prices merely "masked" by "extortionate OPEC prices." Supposedly, Adelman is the source for this conclusion, but again, the source would probably find this an alien context for his original remarks.[16] Similar sorts of evidence gleaned from normally moderate to conservative news sources (such as the *Wall Street Journal* and *Business Week*) are also singled out to illustrate *The Nation*'s writers' anti-oil-industry arguments.[17]

In sum, then, there was a great deal of hoisting the oil companies and sympathizers on their own petards. The result was 23 pro-industry but negatively used sources (13 pro-industry in general, 8 pro-oil-industry and 2 promajor) pitted against 57 strongly argument-supported, anti-industry sources (10 anti-industry in general, 26 anti-oil-industry and 21 specifically antimajor—some of the anti-oil-industry sources probably antimajor as well). Seven source views were neutral, noncommittal, or not expecially relevant to specific arguments.

The attitudes and thoughts expressed in *The Nation* were opinionated, indeed—so much so that some articles were designated opinion/analyses rather than analyses to indicate their extreme biases or intents. The articles were written largely by independent writers who, although they may contribute fairly regularly, are not staff members in the usual sense. Thus, the views presented are their own, which may account for some of the intricacies (and perhaps the overkill) used in arguing special points. However, the accompanying editorials seemed to be in agreement with the spirit, if not the exact details, of the contributors' articles. As a result, *The Nation* received many of extreme attitude/opinion ratings on various topics in its ten articles.

Big business was viewed quite unfavorably three times and unfavorably in two articles (and not discussed at all in the remaining five). Small companies or entrepreneurs, discussed five times, were treated quite favorably once, neutrally four times. Big oil, however, got a unanimous, highly unfavorable reception. Independents were viewed quite favorably three times and neutrally or not at all seven times. The market mechanism, discussed seven times, was greeted very unfavorably six times and unfavorably once. Government action in general was treated seven times—very favorably five times, mildly favorably once, and neutrally once.

Government inaction in controlling multinational oil companies through tax changes and other regulatory measures was criticized severely on three occasions. Nationalization of the oil industry was highly favored twice, and a federal energy company to compete with private firms was advocated quite favorably once. Subpoenaing or requiring information from oil companies was highly favored twice and treated neutrally twice (although in the latter case, the articles complained that information had been slow in coming from the oil majors. An independent research and development agency to gather oil data was highly recommended the three times it was discussed. And action by the State Department to oppose the OPEC cartel was highly favored once, as were a windfall profits tax and rationing to force lower consumption of oil.

The oil industry in general was viewed very unfavorably in nine articles and unfavorably once. The competitiveness of the oil industry, discussed eight times, was denied eight times (twice in relation to pipeline ownership/control). Divestiture was strongly advocated twice, treated mildly favorably twice, and not discussed at all in five articles.

To summarize, *The Nation* discussed the matters of oil industry competition, monopoly, and divestiture very directly and in depth. Its choice of skillful but ax-grinding authors clearly indicate its message—that the oil industry is not now, nor has it ever been, competitive. Its structure and behavior must be altered drastically through government intervention—legislation that will strip it of its special tax privileges, its abilities to control political figures, its inherently anticompetitive integrated form, its accumulated bigness and power. Although government has been the handmaiden of big oil, it must also become the means of breaking up the special interests. It does this, in *The Nation*'s scheme of things, by representing the people and democratic ideals. *The Nation* may suffer inconsistencies, but there is no philosophical wavering, nor does it feel compelled to present the other side. In truth, when reading this publication, one begins to forget what the other side was in the first place.

NATIONAL REVIEW

Between April 1973 and December 1974, when the OAPEC embargo's effects were most prominent and divestiture efforts were beginning to take shape, this weekly opinion publication produced nine articles or nearly eight pages on the energy crisis and charges of anticompetitiveness against the oil industry. All but three of these articles were editorials, and of the three, one was an opinion/

analysis by a regular contributor and two were editorial/analyses. Four were unsigned, three were written by *National Review* editor William F. Buckley, one by associate editor Alan Reynolds, and one by financial columnist/editor William F. Rickenbacker.

Specific topics, in order of frequency of discussion, included oil company profits, criticisms being made of the industry, industry competitiveness, and the media's performance in covering energy matters and government policies. Throughout, the *National Review* took care to defend the industry's contention of competitiveness, expose its detractors, and deplore government interference in the oil marketplace.

Of 45 arguments presented, 3 were pro-industry in general, 29 pro-oil-industry (there was only passing distinction made between majors and independents), 2 anti-oil-industry, and 11 anti-government-regulation policies. The two anti-industry arguments emanated from a discussion of Senator Henry Jackson's contention that the majors had reaped windfall profits and did not pay their fair share of taxes (a position the *Review* obviously did not accept, and said as much after quoting Jackson to illustrate one of Buckley's favorite themes—political demagogy).

The arguments used to illustrate the oil industry's competitiveness were market share or concentration ratios (three times); the fallaciousness of a number of commonly accepted arguments such as squeezing (once), running independents out of business (three times), acting as a monopoly (twice), and price gouging (twice); profits (ten times); oil company shareholders' returns as an indicator of fairly acquired profits (once); the continuing existence of national security considerations in energy policy (twice); and the need for government incentive in the more exotic or risky areas of energy development (twice).

High prices and the bad outlook for energy were generally attributed to government policies that suppressed industry's desire or need to expand domestic oil production. OPEC is blamed once, but is treated as too obvious a cause to dwell on. More at issue in *National Review* eyes are limitations on production extant in some states (twice); the no-longer-necessary depletion allowance (once), tariff (once) and import quota formulas (three times); price controls (three times); the ecological lobby, which has discouraged offshore oil production or other ventures (once); and too much overall government involvement in private enterprise affairs (twice).

In an opinion/analysis article, "Monopolies and Antitrust," financial editor-publisher and *National Review* columnist William Rickenbacker discusses what he (and five antitrust writers he quoted) thought concentration ratios meant. His conclusion is that they mean different things in different contexts and one must have expert knowledge of the case in point. "Concentration ratios say nothing about competition within an industry. An increase in concentration might increase competition. [One] must know how and where firms operate."[18] His own definition of a monopoly is an "unholy alliance between big government and big business," but he cites those who have studied concentration and industrial organization and finds that "There is little evidence that competition has given way to concentration and monopoly over the years."[19]

Another article, an editorial, points out that the oil industry's ranking among major industries is not very impressive—number seven out of the ten largest industries.[20] And an editorial/analysis, on commonly held misconceptions about energy economics, argues that oil companies are neither unduly concentrated nor anticompetitive.

> Economic theory tells us that a shared monopoly will produce somewhat less than a competitive industry, and at a somewhat higher price. . . . Having found the wealth-maximizing combination of supply and price, the trend over time is the same in competitive or monopolized industry—both produce more as demand and price rise. To say that oil companies cut supplies to raise prices therefore implies that the monopoly is brand new. If they had the power before they would have used it before. . . . And the related notion, that the oil companies are using high prices to "squeeze out" competitors makes no sense at all.[21]

A big area of concern for the *National Review* is the (unwarranted, it feels) hullabaloo about prices and profits. Repeatedly, it complains that the media and politicians have ignored or confused the facts. One editorial looked at two publications' pronouncements during the same week in the fall of 1974. The New York *Times* (said to have selected figures carefully, thus misrepresenting the facts) claimed that companies had cut production and raised prices. At the same time, *Newsweek* said that the majors, by not cutting back prices, were instigating price wars and running independents out of business. Huffs the article, "So it really doesn't matter what the oil companies do. . . . They are still going to be scolded."[22] Another editorial, blasting Henry Jackson for his statement that multinationals were earning obscene profits plus not paying their fair share of taxes, argues that prior to 1972 profits were below industry average; that during the embargo they rose temporarily as a result of price increases; but that they are [as of late 1974] "less obscene" and, in fact, look as if they will continue a downward trend.[23] One editorial points out that oil company profits did not actually add much to the price of oil, whereas OPEC price demands did.[24] And yet another argues that if profits had been good, oil stock shareholders would have done better on their investments than they did.[25]

As for government policies, especially those that restrict the free market, the *National Review* has few good words. However, this philosophy produces some fairly complicated (and perhaps surprising to the uninitiated) observations by the publication's editor, William F. Buckley. He opposes controls but he also opposes privileges that put the market out of equilibrium by favoring certain segments. For example, he objects to "the oil producing states [using] the multinational oil companies as, in effect, tax collectors."[26] That is, the companies, rather than buying oil from the states, act directly as producers and thus pay the states lease bonuses or royalties which increase costs, benefit only certain states, and raise prices. Similarly, he dislikes the "exclusive" benefits that quotas, the depletion allowance, and tariffs have wrought.

This same hands-off philosophy is evident in several editorials decrying price controls (as discouraging oil exploration and development or alternative energy development by private enterprise) and "conservation" measures that limit production. One cites industrial organization economist Paul MacAvoy who wrote in a scholarly paper that ". . . state conservation efforts have limited production capacity, while federal policy has limited price, with the result that the country is in the position of excessively expanding imports."[27] Another measure, criticized as "pork barrel socialism" is the Federal Oil and Gas Company (FOGC) concept. One editorial objected to the government subsidies (tax exemptions plus a choice of 20 percent of federal lands offered for lease) that the proposed FOGC would receive. The article chastized *New Republic* writers Peter Barnes and Derek Shearer for their irresponsibility in advocating such a plan.[28]

Somewhat paradoxically, there was one aspect of government intervention favored by the *National Review*. This was the protection of "national security," that is, the willingness of government to support (or at least not obstruct) capital investment in certain energy areas. Writes Buckley:

> During the Second World War, faced with the interdiction of our rubber supplies, the government developed a synthetic rubber, and the private companies then produced it, competing with each other. The next great breakthrough, before the efficient use of atomic energy, is probably in the direction of the liquefaction of coal. This, it is thought, will require a capital investment outside the reach of the oil or coal companies. They talk of $20 to $30 billion—about what it required to reach the moon. Other reforms should be made quickly, easing the government out of the way of private initiative. But the government must stand by to meet the needs of the national security.[29]

A pervading argument, seen in earlier quotes, is that some members of Congress and the media have either used the energy crisis for personal aggrandizement or exploited the opportunity to keep the public in the dark about "basic axioms of economics, about supply and demand, and resource allocation and the function of price rises." Television becomes a primary target of Buckley's consternation.

> It is the impact of television journalism, and there isn't anyone in the United States who denies its importance, primarily in estab-

lishing a public mood. It is by the requirement of the medium, anti-thought. It is at the service of those who express readily communic-able attitudes and emotions: indignation, despair, joy, skepticism, hatred. In the colloquy in question we have a politician saying the obvious things, but those things that he was saying stir the resent-ments, latent and matured, of everyone who deplores a) paying more for gas or fuel, or b) finding it hard to get gas or fuel. Roughly every-body. By contrast, the oil executive is speaking about such abstrac-tions as demand, supply, costs, controls, allocations: so that the anchorman gives him very short shrift.[30]

National Review's 40 sources (23 cited, 17 mentioned), used to bolster its mainly editorial opinions, were composed largely of independent or academic figures (17 references for 9 individuals) and media sources (14 references for 5 individuals). The remaining source references were to American Petroleum In-stitute figures (2), members of or studies by the U.S. congress (4) and a state congressman (3). No specific oil industry or government agency sources were mentioned. The sources who represented anti-industry views were treated nega-tively; that is, if their views were stated, they were simultaneously refuted. How-ever, since this occurred only in editorials, the criticism of sources was made fairly openly. In analysis articles the problem of contrary views was overcome, quite simply, by ignoring them. It does appear, however, that independent or academic source views were presented without distorting the intent of their admittedly selected quotes. And, in some cases, their complicated arguments might have been elaborated upon too much for the nontheoretically oriented reader.

The attitudes and opinions expressed in the *National Review* were gen-erally probusiness and antigovernment. Business in general was treated quite favorably in two articles and favorably in another (and neutrally or not at all in the rest). Big business was treated favorably in one article whereas small busi-nesses were regarded neutrally (four times) or not at all (5 times). Big oil re-ceived a very favorable boost twice and a somewhat favorable once, but was treated neutrally the other six times. Independents were not really discussed.

The market mechanism, ignored in only one article, was viewed quite favorably in all the rest; and response to government action in general paralleled the promarket response in the opposite direction. The only kind of specific government action highly favored (twice) was protection of the U.S. national security (threatened by oil interruptions). All other specific government action plans were flatly rejected: FOGC-type plans (twice), price controls (four times), oil import quotas (three times), oil production quotas (once), the depletion al-lowance (once), windfall profits tax (once), tariffs (once) and a gasoline tax (once).

The oil industry as a whole was regarded quite favorably six times, mildly favorably once, and neutrally or not at all twice. The view that the oil industry is competitive was highly favored four times, faovred twice, mildly favored once, and considered not applicable once. Divestiture, no doubt considered absurd even to suggest, was not discussed in any of the nine articles.

In summary, the *National Review*, in discussing oil industry competitiveness and proposed or existing government policies for dealing with the energy problem, espoused a conservative concern with too restrictive government and, in fact, objected to any measures that gave special benefits to particular groups. It was highly critical of the media's and Congress's performance in translating energy economic issues to the public, calling them demagogues. Certainly, the *National Review* lapsed into some of its own pedantry and was not the most sympathetic of translators. But it did present some interesting sources (an exceptionally large proportion of independent or academic economic experts) and provided a well-informed, relatively consistent set of arguments. Unfortunately, in this rare atmosphere of purist conservatism, there was little consideration of whether some reforms or monitoring functions might not be in order, so that oil economic activities could be made more accessible to the public, and further, whether a fully unfettered oil industry would serve what the public (in its ignorance?) perceived to be a common good.

THE NEW REPUBLIC

This weekly opinion publication covered the oil monopoly/divestiture issue at length and in depth—19 articles, 31 pages—from June 1973 through December 1975. Six of the articles, editorials, were unsigned. The remaining 13 articles were 10 analyses, 2 editorial/analyses, and 1 book review by three *New Republic* staff members or contributing editors and nine outside authors (including one or two fairly regular contributors to the magazine).

The staff writers were Eliot Marshall, then associate editor specializing in energy matters (two articles); Melville J. Ulmer, a contributing editor (two articles); and Peter Barnes, West Coast correspondent, writing one article with Derek Shearer. The outside authors were: Muriel Allen, *Journal of Commerce* writer collaborating on one article with Richard Levy, a Washington, D.C. lawyer and oil consultant; Phillip M. Stern, contributor (one article); Robert W. Dietsch, contributor (two articles); Derek Shearer (co-authoring one article with Peter Barnes), consultant to the Exploratory Project on Economic Alternatives; Josiah Lee Auspitz, current affairs writer and president of the Ripon Society (one article); Charles L. Schultz, former director of the Bureau of the Budget (one article); Daniel Yersin, a writer on international oil matters (one article); and M. A. Adelman, professor of economics at M.I.T. (one article).

The topics in *The New Republic*, in terms of frequency of appearance, included divestiture (five times—once as the subject of an editorial, four times in conjunction with other topics); the competitiveness of the oil industry and related antitrust violations or measures (four times); the reasons for shortages, profits, thwarting the OPEC cartel, and the lack of data about the oil industry (three times each); prices, price decontrol, and federal energy company (FOGC) proposals (two times each).

The arguments incorporated in *The New Republic* were numerous and largely critical of the major oil companies, but also critical of the OPEC cartel (with which the oil companies were suspected of having colluded or cooperated to maintain anticompetitive control of the petroleum market). The government, particularly in the area of taxation policy, also came under criticism in approximately one-fifth of the arguments. Of a total of 227 arguments, 169 were anti-major, or less often, pro-independent, 25 were pro-oil-industry, 31 were opposed to various government policies or actions, and 2 were progovernment. (This latter figure means that only two arguments supported current government policies; many arguments implied proposals for or changes in existing federal policy.)

The most frequently discussed indication of industry noncompetitiveness was cooperation or collusion among majors within the industry, a concerted effort to maintain exclusive control—17 times. This was followed by squeezing independents or driving them out of business (16); excessive or windfall profits (15); acting as a monopoly or oligopoly, especially in the area of pipeline control (13); restricting supplies by holding down either production and distribution (11) or refinery capacity (1)—an argument related to monopoly control over the means of production and squeezing; failure or unwillingness to disclose information or lack of independently gathered data (7); dealing into the hands of OPEC (7); taking advantage of governmental policy or exerting too much political influence or power (5); using accounting devices or fund transfers to obscure the true profit picture (4); contriving (1) or taking advantage of (3) shortages; engaging in deceptive advertising (2); and using mergers or acquisition to reduce competition (1). In addition, a federal energy company plan was proffered (7 times) as a way of injecting competition into the oil industry and serving as a yardstick for pricing and supply decisions.

High prices and shortages were attributed in *The New Republic* to a combination of major oil company, OPEC, and government policies or actions. The major or multinational oil companies were considered responsible for high prices and shortages 16 times, or to have acted in concert with OPEC 4 times, or within the context of pro-oil government policies that they helped to influence 27 times. Government policies singled out as especially onerous were tax loopholes or advantages (10) including the depletion allowance (5), foreign tax credits (1) and intangible drilling expensing (1); import quotas (3); price control deregulation (2); and the relaxation of environmental standards (3). In addition, there were many antigovernment-policy arguments that criticized government policies or inaction independent of any oil company influence or manipulation. The move toward price decontrol, the failure to devise a system for studying the industry and for taxing windfall profits, and last, but foremost, the failure to get at the root of the majors' domination of the oil market—their hold over the means of production and their integrated structure—were the primary government faults that concerned *The New Republic*. Following are some quotations to illustrate the predominant themes in this publication's major oil company/ government criticisms.

Charges of collusion among major, integrated companies—the restriction of supplies, price fixing, or the failure to avert or warn of impending shortages—began to be heard in state antitrust suits in 1972-73. In June 1973, *The New Republic* began commenting upon the reasons for oil shortages. One article, an editorial/analysis, cites the usual explanations for shortages offered by majors, "not a disinterested source": worldwide shortages of sweet crude, insufficient refinery capacity, jump in demand because of strict but fuel-inefficient new air standards and the recently suspended oil import quotas. But this does not satisfy the article's author, Eliot Marshall, who writes, "The climate is right for conspiracy theories, but so far the government has not found the case strong enough, or the public angry enough to take the oil companies to task."[31] In another article, Robert W. Dietsch, hailing the efforts of state attorneys general in oil antitrust suits, quotes extensively from a report commissioned by the National District Attorneys Association, the Leonard Report:

> ". . . The common threads that prevade the oil industry are knowledge of each other's activities through a complex series of corporate agreements in joint bidding leases, joint production, joint ownership of pipelines, crude oil exchanges, refined product exchanges and possible sales or exchanges of marketing, transportation or refining facilities; identical courses of action relating to pricing in gas war situations; and series of interlocking primary and secondary relationships with the nation's largest banks and insurance companies. The majors also have acquired smaller producers and refiners through purchase or merger and have, by these actions, caused an apparently catastrophic effect on the independent retail marketers who were through highly efficient operations and on a retail price competition basis making significant inroads into the retail market."[32]

An analysis article by Melville J. Ulmer perceives collusive behavior not only among majors but between the big oil companies and OPEC: ". . . today the American public is caught in a pincer movement between a domestic cartel and a foreign cartel, essentially two arms of the same body."[33] And Josiah Lee Auspitz argues similarly that the U.S. companies had initiated a policy of multinational and Arab cooperation as early as 1945, a policy it tacitly continued through inaction.

> The oil companies, hardly philanthropic organizations, saw the writing on the wall. The 25-year harmony of corporate and government

interests was coming to an end. The Anglo-American military presence was less reliable, public opinion was dead set against them, the oil producers were getting restive, and the U.S. position on the Arab-Israeli conflict could only work against them. If their position was untenable, they might as well get out comfortably. Hence they did nothing to oppose measures by OPEC to jack up the price of oil to a level that now makes it possible for them to sell their foreign holdings on lucrative terms. And the US Departments of State and Treasury, accustomed to taking cues from oil company officials, responded to the most outrageous moves by the OPEC countries only with admonitions.[34]

Another antioil argument, relating to the majors' ability to control crude oil, is that of squeezing independents. This appears in several articles. In one analysis, Ulmer writes that the depletion allowance greatly aggravated the independents' plight:

The oil industry has been carefully planned, but in the interest of monopoly profits not the public. For example, the 22 percent depletion allowance is calculated on gross income from *oil production*, not refining. This arrangement stimulates the integrated producers to maximize their tax saving by keeping prices of crude oil at the wellhead abnormally high . . . it also yields another significant contribution to their financial welfare. For the majors sell part of their crude oil to independent, nonintegrated refiners. By hiking its price they can squeeze these potential competitors at will, narrowing their profit margins until they cry "help!" Help means selling out or merger. Between 1950 and 1967, according to the Federal Trade Commission, the majors in this way absorbed 73 large independents with aggregate assets of more than six billion dollars.[35]

The New Republic also took special pains to examine one aspect of oil operations that it thought was especially monopolized by major oil companies—pipelines. One article ran an exchange between two congressmen on opposite sides of the pipeline ownership issue:

A conversation between Senator Haskell, a strong proponent of tough antitrust enforcement, and Senator Hansen, an advocate of the Alaska group's pipeline proposal, sums up the matter. Haskell asked, "Why can't we get an independent carrier for Alaskan oil?" Hansen replied, "Because there's so much risk involved." "What risk?" Haskell asked. "With a 10 billion barrel field and an exclusive market? Give me a franchise and I'll resign from the Senate and run the operation."[36]

The subjects of prices and profits become intertwined in a number of articles concerned with the monopolistic or oligopolistic nature of the major oil companies. One editorial complained in early 1974:

> Unfettered free enterprise is having a field day, and the government chooses not to challenge the monopoly that keeps prices rising. . . .
>
> This [profits] would cause less concern if there were a spark of competition left among the major producers. There isn't. Nor would there be so much anxiety if the energy bureaucrats would regulate prices strictly according to cost. They don't.[37]

And in early 1975 another editorial, trying to beat back price decontrols, questioned how there could be such great differences between production costs and prices.

> The urgent campaign to decontrol oil ignores some tough questions that have come up in the years since the embargo. For example, people wonder why Arab oil that costs only 15 cents a barrel to produce sells on the President's "free market" for $11 a barrel. Why does American oil that a couple of years ago sold for three or four dollars a barrel sell for as much as $11 a barrel today? The cost of producing it hasn't gone up that much.[38]

A third editorial on price decontrol, lamenting the President's veto of a bill extending price ceilings on old (pre-1973) crude oil, comments that "Ford's program will allow the companies a wider profit margin and a hope of eliminating all regulations sometime in the distant future."[39]

Another area of criticism, or at least unease, was what *The New Republic* called the "oil information shortage." One editorial complains that the Federal Energy Office received little information on how bad the shortage was and "the information it does have comes from the oil industry."[40] It also quotes several senators and representatives either dismayed by an inability to obtain reliable data or working on a bill to create a federal energy data office. (The article does note that several major companies complied in part with data requests). In addition to lack of data, there was felt to be lack of a model or yardstick by which to measure private industry's performance. Thus, several articles explored (mostly in laudatory tones) proposals for a federal energy company (FOGC). Acknowledging one of the American Petroleum Institute's objections to a public oil and gas company, one editorial goes on to say:

> Ikard [API president] did make a good point that's worth repeating: he said the FOGC will not alleviate the present energy shortage. A corporation launched this year would contribute very little to oil production in this decade. FOGC would, however, give the govern-

ment direct and dependable data on the drilling, pumping, refining and marketing of oil and gas. It would do extensive research on fuel that lies on government property, filling in some of the gaps in the government's knowledge of its own reserves. It would encourage competition not just by trying to undersell the majors, but by seeking out independent markets for its crude oil. Though the FOGC would not make fuel available at 1973 prices, it would provide an independent standard by which to judge the majors' performance. Congress ought to get behind it.[41]

A great many of the anti-industry arguments in *The New Republic* revolve around the issue of power, contending that the major oil companies, with or without OPEC's input, have accumulated too much influence and have gone too long without careful government scrutiny, regulation and possibly even structural reorganization. As Ulmer writes in "The Oil Masters":

We resent the Arabian producers because of the inconveniences and high fuel prices we are suffering, but hardly a word is heard about the international cartel that has so carefully regulated the world's supply of petroleum, dividing markets and fixing prices, ever since the first and infamous Achnacarry Agreement of 1928. On the contrary, the administration, Congress and the public look for succor to our corporate petroleum giants, the leaders of that cartel. . . .

Wrath against the Arabs for international blackmail is not unjustified, nor is the new US "war of independence" to loosen us from reliance on foreign sources unreasonable or quixotic. But both mask the fact that the United States can never be secure in its supply of energy until it controls the controllers. Those controllers—Exxon, Standard Oil of California, Texaco, Mobil, Gulf and a few other "major" producers—rank high among the 100 largest industrial corporations in the United States. They have acquired such enormous wealth and wield such awesome power, enough to strike fear in the most courageous of statesmen, that when necessary they can mobilize the machinery of government in the service of their balance sheets. Franklin D. Roosevelt once said, "The trouble with this country is that you can't win an election without the oil bloc and you can't govern with it."[42]

Another editorial comments on the industry's above-the-law attitude:

The government has from time to time taken shots at the anti-competitive structure of big oil, and it continues to process lawsuits and restraint-of-trade inquiries. . . . Congress could pass new antitrust legislation, or it could simply order a divestiture by big oil of pipelines and refineries. That would help. But Congress has shown no enthusiasm for such drastic action. Nor is there any guarantee that oil corporations wouldn't find new ways to work their will within the literal restrictions of the law while ignoring its intent, as they do to-day.[43]

And finally, an editorial—with some reluctance—advocates divestiture because:

> Since the onset of serious price gouging in 1973, it's become apparent that unless the government intervenes to disrupt the private crude oil barter system in which all big companies participate, there never will be a free market. On the contrary, without some sort of policing, the industry is likely to become even less competitive and more dominated by the largest conglomerates than it is now. . . .
>
>   It's also apparent that the agencies that should enforce the law are unwilling to attack collusion in the oil industry. The federal regulatory commissions under Nixon and Ford set themselves the goal of increasing energy prices. The Justice Department is silent, and even the President dreams of other things, such as how to distribute new subsidies to makers of synthetic fuel.[44]

There are in *The New Republic* some mediating, qualifying, or balancing statements to the effect that the oil industry has been competitive, willing to modify its own behavior, or in favor of some governmental regulatory action. The article on the information shortage, for example, names the companies that had provided some figures fairly quickly on the prospects of worsening shortages.[45] And one of the FOGC articles does accede to industry's argument that such a program would do nothing to alleviate current shortages.[46] An article on profits quotes a major oil head as saying that the depletion allowance is no longer necessary, an indication of industry's possible potential for self-reform.[47] *The New Republic* was also careful to point out, at several junctures, that the charges of an oil conspiracy had never really been substantiated (although some authors, nonetheless, accepted collusion as fact).

There was one area in which this publication came close to industry's point of view. The oil companies were thought, in many instances, to have assisted OPEC in setting prices and limiting production. However, the OPEC cartel was blamed for high prices and/or shortages 13 times, almost as often as major oil companies (17 times), and several articles tackled the problem of how to thwart or weaken the OPEC cartel. The effect of this desire created something of a dilemma for several writers because it meant that some domestic oil development had to be encouraged. (But how to do so without giving the majors free rein to acquire greater power?) Ulmer's solution was either a FOGC-type proposal or making the oil companies quasi-public utilities, subject to full disclosure, government overseeing, production quotas, tax penalties, price controls, and so on. Other writers like Auspitz and Yergin preferred a diplomacy route—increasing relations with non-OPEC (but potentially oil-rich) nations like Great

Britain or some of the non-Arab oil producers, or encouraging greater conservation (persuasion, or if that did not work, gasoline taxes or, more favored, rationing).

Finally, M. A. Adelman, in an uncomplimentary review of Christopher Rand's *Making Democracy Safe for Oil*, suggested that Rand's priorities were all wrong. He should, argued Adelman, be concentrating on diminishing OPEC's power, not punishing the oil companies that have no control over OPEC pricing. Adelman, who has long objected to oil company tax, import, and production privileges, may have surprised *The New Republic* editors when he wrote: "Our troubles are due not to villany but to muddle. Rand, by making villains of the oil companies, and successful villains at that, has made the muddle a bit worse.[48]

The predominant sources used in *New Republic* articles were government sources—68 senators or representatives, 39 federal officials or agencies, 13 White House sources and 2 state-level officials. (This was out of a total of 172 sources.) Of the 68 congressional source references (only 14 of which were cited directly), 49 opposed current oil industry behavior or structure. The most frequently cited congressional source was Senator Frank Church (2). Most often mentioned were Senators Jackson (10), Hart (7) and FOGC proposer Stevenson (6). The federal official sources most often cited or mentioned were the FTC (10 times) and the FEA (7 times). The total number of government source references was 122.

Industry sources included 11 majors, 1 independent and 5 American Petroleum Institute references for a total of 17. News sources totaled 8, unnamed industry critics (6), and OPEC sources critical of U.S. oil companies (2). There were three incidental sources.

Considering the strong pattern of anti-industry arguments that appeared in *The New Republic*, its distribution of sources included a reasonable proportion of sources that were either pro-industry or opposed to government regulatory or antitrust measures. There were 33 pro-industry sources, 11 specifically promajor, 1 specifically pro-big-business, and 6 anti-government-policies, for a total of 51. Conversely, there were 57 anti-industry sources, 46 specifically antimajor or pro-independent, and 13 specifically favoring government policies that would curb the excesses of big business, for a total of 116. In addition, there were eight sources whose statements or views led to neutral or nonpertinent conclusions. Thus, anti-industry outnumbered pro-industry sources more than two to one.

The amount of spatial coverage given different sources showed an even greater disparity. Of 19 articles, 1 gave greater space to anti-big-business sources (an article on the oil and auto industries), 5 gave greater space to anti-oil-industry sources (with the emphasis, however, on anti-big-oil) and 6 favored sources whose criticisms were specifically directed against majors—a total of 12 articles. Only one article (Adelman's rejection of many of Rand's conclusions) could be said to favor pro-industry—or at least not anti-industry—source views. Nearly a third, or six of the articles, did a good job of providing source space balance.

The attitudes and opinions expressed in *The New Republic*, while having a basically liberal philosophical cohesion, vary in particulars. Business in general, when alluded to at all, was treated neutrally in two articles and mildly favorably in two. Big business was treated very unfavorably twice, somewhat unfavorably twice, and neutrally twice. Big oil was viewed quite unfavorably in 14 articles, unfavorably in 2, somewhat unfavorably in 2, and mildly favorably in 1. Independents, although not cited or mentioned as sources, were given frequent consideration. They were depicted very favorably in eight articles, favorably to partially favorably in three, and neutrally in the rest.

The market mechanism was viewed as a very unsatisfactory means of producing competition in ten articles and was greeted mildly unfavorably in four, neutrally in four, and favorably in one. Nonspecified government action was viewed highly favorably 11 times, favorably three times, mildly favorably once, neutrally three times, and mildly unfavorably once. Specific kinds of government involvement or actions that were strongly endorsed most often (in anywhere from three to seven articles) included promoting conservation, prodding Congress to act on energy legislation, removing government privileges granted to oil companies, a FOGC-type plan, divestiture, antitrust actions other than divestiture, compelling information disclosure by oil companies, price controls, and allocation formulas. Other types of government responses endorsed infrequently (only once or twice each) included tax subsidies to spur domestic exploration, price decontrols, maximum import and production quotas, abolishing the FEA, and a gasoline use tax coupled with offsetting tax cuts for the poor.

The oil industry as a whole was treated very unfavorably in three articles, unfavorably in six, and somewhat unfavorably in four (neutrally in the remaining four). The belief that the oil companies had behaved competitively was supported very favorably in only one article. Thirteen indicated strongly that it was not competitive, one was mildly of this opinion, and four were neutral. An interesting aspect of *The New Republic's* approach was that it advocated various degrees of divestiture. This distinction indicated that it had given some practical as well as philosophical thought to what areas of the industry were least competitive and should be attacked first (in the event an overall divestiture move failed). Two articles highly favored pipeline divestiture, one highly favored separation of refining operations from integrated firms, and one highly favored both horizontal and vertical divestiture (although its greatest concern was with vertical). The remaining 14 articles discussed oil company abuses and problems but not divestiture per se.

In summary, *The New Republic* engaged in a lot of examination and soul searching about the oil monopoly question and federal means of either creating competition or dismantling the present oil industry structure to allow for competition. Its favored option was tighter governmental regulation of companies to monitor information and halt anticompetitive practices, coupled with a federal energy company to act as a yardstick and competitor to private industry. It dropped this option rather reluctantly (or rather the option melted away due to congressional coolness toward it) and went on to endorse divestiture (but none too heartily). *The New Republic*, while maintaining its biases, still provided some thoughtful discussion, including views with which its editors did not agree. A central premise in its thinking seemed to be that radical departures from the status quo may not be ideal solutions, but past nonsolutions or inaction in righting wrongs have produced a stagnating system. Thus, although there was a tendency for *The New Republic* to leap to preconceived conclusions, a lot of thinking seems to have preceded the jump.

NEW YORK TIMES MAGAZINE

This Sunday section of the New York *Times* featured 6 articles or approximately 23 pages on oil company behavior, the energy crisis, and the technological and socioeconomic factors involved in the production and sale of oil from December 1973 through December 1974. Three of these were analyses by Brit Hume, *ABC News* Washington-based investigative reporter; Robert W. Stock, a senior editor for the *Times Magazine*; and Paul Lewis, Washington correspondent for the London *Financial Times*. One was a satire on shortages by *Times Magazine* columnist Russell Baker. And two were letter-to-the-editor-type exchanges. The first was between Alan Reynolds, attacking many of Hume's conclusions, and Brit Hume, with additional comments by Frederick C. Thayer, associate professor at the University of Pittsburgh, who agreed with Hume. The second letter battle occurred between Maurice F. Granville, Texaco's chairman, and Brit Hume (with a reply and counterreply to Hume's original article and exchange with Reynolds).

The article most pertinent to oil industry behavior and structure and the monopoly question was Hume's. Two months after the embargo, in "The Case Against Big Oil (Why was there a shortage before the shortage?)," Hume traces the history of a number of big oil's transgressions. Thus, many of the anti-oil-industry arguments appearing in the *Times Magazine* sprang from this article. Of the 78 arguments incorporated in this publication, 44 are antimajor and 18 are either anti-oil-industry or opposed to government policy actions toward the industry—a total of 62. On the other side of the issue there are 14 pro-industry and 2 specifically promajor arguments—a total of 16 pro-oil arguments.

The anti-oil applied most frequently argument was that the major integrated companies had clamped the lid on supplies—a deliberately contrived shortage that resulted in squeezing of independents, higher prices, and petroleum product scarcities. The majors were said to have slowed production (3 times), held up expansion of refinery capacity (8), withheld supplies from the market (4) and shared oil or refined products among themselves (3)—a total of 18 antimajor supply arguments. Other arguments included squeezing or running independents out of business (7), collusion among majors (2), acting as a monopoly (3), under- or misreporting reserves or using spurious accounting techniques (3), earning windfall profits (4), not providing enough information to Congress (1), and using inappropriate arguments (such as national security) or scapegoating others (for example, OPEC) to try to deceive the public (4).

Big oil's influence wielding in government was also a prominent anti-industry claim. While OPEC was blamed for high prices and shortages three times, the major oil companies were blamed twice as often and government policies that aided them were thought responsible eight times. Citing some of the spoils acquired by the majors, Hume writes:

> . . . the industry's giants have failed to get the price increases they have longed for. But they have accomplished a number of other objectives. The Alaska pipeline has been cleared for construction by Congress and environmental roadblocks—to offshore drilling, for example—seen about to fall. The pesky independent companies have been gravely weakened. And the majors' sagging profit picture of the late nineteen-sixties and early seventies has taken a dramatic turn for the better. . . . There are some who believe the major oil companies conspired to bring about these results. Indeed, the Federal Trade Commission . . . has filed a huge antitrust case against the top eight oil companies accusing them of collusion to monopolize the industry.[49]

And British writer Paul Lewis perceives the oil industry (and the big business ethos in the United States) as having worked with the cooperation of the government to control oil's price and supply.

> What is true, however, is that since the invention of the internal combustion engine, the industrial powers have sought to establish and maintain a monopoly of the world's oil supplies and that the pious horror with which Americans react to the monopoly practices of the O.P.E.C. cartel today is quite unjustified. It is from the West that its members have learned about cartelization.
>
> From the very beginning of the automotive age, the British and American governments have worked hand-in-hand with the major oil companies to ensure the industrial world a steady supply of oil at stable prices.[50]

A satire by Washington humorist Russell Baker, beneath the fluff, also suggests that the big oil companies had conspired to create the "Crisis Crisis": "The world needed a great new crisis. . . . If my theory is right, the oil people all met in the crisis room when the latest Arab-Israeli war began."[51] The oil people's ideas for creating "total credibility breakdown and paranoia," include scapegoating the Arabs and the Shah of Iran, keeping loaded tankers offshore until prices go up, closing gas stations and saying there is no more oil, and asking the citizenry to make sacrifices so the sellers can all go ahead and raise prices.

Adds Baker: ". . . the beauty part, of course, is that everybody gets even richer, except people who weren't rich to begin with, so nobody suffers."[52]

But there are also some counterarguments to such contentions. Reacting to Hume's analysis, *National Review* associate editor Alan Reynolds writes:

> Aside from some annoying innuendoes, Brit Hume's "The Case against Big Oil" (Dec. 9) was no such thing. It was simply a very selectively translated version of the Senate staff study mentioned in the first paragraph. . . . Refineries did not operate at top capacity for a few months because of mechanical repairs, or because they simply didn't have enough crude oil to refine. Imported oil was rising in price very rapidly and could not be refined and marketed profitably under Phase II controls. . . . By keeping the prices of petroleum products artificially low, the Government has encouraged waste and discouraged production. The resulting shortage, like the beef shortage, will worsen until prices are permitted to bring supply and demand into balance.[53]

Hume retorts that:

> Mr. Reynolds makes a concise summary of the oil industry's arguments. . . . His basic thesis—that the oil companies would have produced more if it had been profitable to do so—deserves some comment. Mr. Reynolds seems to have forgotten that this is the industry that has long enjoyed a vast array of tax benefits and production and import restrictions. These are all artificial devices which interfere with the free-market mechanisms to which economic conservatives like Mr. Reynolds are supposedly devoted.[54]

Hume is supported somewhat in his supply monopoly argument by the University of Pittsburgh's Frederick C. Thayer (his field was not indicated), who is worried about the planning elements of gasoline distribution or allocation. Thayer writes: "Shortages require that we search for something we have avoided—the planning of distribution within a scheme that is something older than monopoly."[55]

Five months after Hume's article originally appeared, the magazine also ran a letter from Texaco chairman Maurice Granville. He countered Hume's statements that supplies had been deliberately kept from the market and cited several figures about which he claimed Hume was mistaken. Granville charged that Hume had selected out information to support his premises, ignoring contrary data (for example, a Federal Power Commission study, a Treasury Department report, and crude oil stock figures).[56] Hume was given an opportunity to reply and reasserted some of his original conclusions, citing studies other than the ones Granville had mentioned.

Senior Editor Robert W. Stock's largely technical and descriptive article on the journey of a barrel of oil from Conroe, Texas to the pump offers some explanations as to why shortages had occurred. Among these were the high costs of expanding refinery capacity while meeting environmental standards, the limitations on imports (once but not now to industry's advantage), government disincentives to expanded production or refining (such as price ceilings, market demand prorationing), and the fact that domestic production had become increasingly technical and costly. Writes Stock: "Thus, in this troubled month of February, 1974, the refineries are doing their limited best and gasoline is in short supply."[57] So, while this author says that he does not think industry should be spared criticism, he agrees that the matter of energy adequacy has become quite complex and that industry is competitive and is seeking to alleviate supply problems. In a companion snippet (not really a separate article), Stock also interviews Exxon's Keith Jamieson who "seldom dodged a question." With some admiration but with a reserve of skepticism, Stock writes:

> When Jamieson says, "We're a corporation," that tells it all. The odyssey of the barrel of Conroe crude, for all the political and social ramifications along the way, is not ordered and controlled by a charitable institution. Exxon and its fellow majors are devoted first and foremost to the proposition that profit is their most important product, with service to the public weal as an assumed byproduct. That assumption deserves—and is beginning to receive—the closest scrutiny.[58]

The sources used in the *Times Magazine* to support anti-industry views comprised about two-thirds of the total of 27 sources (19 cited and 8 mentioned). Twelve of these were government sources. Three references were to the Nixon administration and nine were to various commissions and studies: the FTC, the Office of Emergency Preparedness, the Treasury Department, the Senate Permanent Investigations Subcommittee, the Texas Railroad Commission (responsible for the prorationing program), the Federal Power Commission, the U.S. Cabinet Task Force on Oil Imports, and the Interior Department's Office of Oil and Natural Gas.

The nine major oil company source references included quotes by two oil executives and the API. There was also a citing of figures from an industry publication, the *Oil and Gas Journal*. There were no independent or academic consulting sources (other than Frederick Thayer's use of himself as an academic authority). In sum, there were 7 pro-industry and 3 specifically promajor sources—for a total of 10—and 5 anti-industry (but with emphasis on large company abuses) and 12 specifically antimajor sources—for a total of 17.

Source space was fairly evenly divided, tipping slightly toward anti-oil sources. One article favored anti-industry sources (Lewis's article on the effects of shortages on underdeveloped nations); two featured antimajor sources (Hume's and Baker's—although Baker really used no sources other than his wit); one gave greater space to pro-industry sources (Stock's article) and the two "letter battles" divided space fairly evenly between pro and anti writers (but gave Hume a bit of an edge through opportunities to defend his article and through the inclusion of Thayer).

The attitudes and opinions expressed in the *Times Magazine* varied along lines similar to argument and source coverage. Because two of the articles actually contained five letters or mini-articles, the attitude/opinion ratings have been applied to nine rather than six articles. For example, big oil was treated very favorably in one article (Granville's), favorably in one (Reynolds's), mildly favorably in one (Stock's), very unfavorably in Hume's three articles and in Baker's, and somewhat unfavorably in Lewis's (which, although critical of big oil had more pressing points to make). Independents were viewed very favorably in three articles, favorably once, mildly favorably once, and neutrally in the rest.

The market mechanism was treated very favorably in three articles, quite unfavorably in four, and neutrally in one. The intervention of government in oil matters was highly favored once, favored once, highly disfavored once, and addressed neutrally or not at all in the rest. Specific forms of government action favored or rejected were government tax privileges granted to oil companies— very unfavorable (twice); government taking the initiative in international settlement of the OPEC problem—highly favorable (once); and government disincentives for expanded production—very unfavorable (once).

The oil industry as a whole was depicted very favorably twice, favorably once, very unfavorably four times, unfavorably once, and neutrally once. The competitiveness of the oil industry was highly accepted in three articles, strongly rejected in five, and treated neutrally in one. There was no clear-cut discussion of divestiture in these *Times Magazine* articles, but some form of breakup (mentioned mostly in relation to pipeline and refinery ownership) was highly favored twice, somewhat favored once, mildly favored once, highly disfavored once, and treated neutrally in the remaining articles.

In summary, the *New York Times Magazine*, from late 1973 through late 1974 focused most of its attention on shortage and supply issues. In fact, its spurt of oil writing productivity coincided with the temporary swell in public panic about oil, then subsided. Thus, a great many questions were neither raised nor answered by this publication. Despite the quantity of space devoted to this subject (nearly 23 pages), the *Times Magazine* was actually rather low on argument content and incorporated only a sprinkling of sources whose views clustered repetitively around the same themes. What coverage the magazine did afford was of two disparate varieties. One type was of the hard hitting investigative school. The other used a plodding but thorough method of examination and provided informative but noncontroversial insights into the petroleum industry. The juxtaposition of these articles was interesting while it lasted, but neither side received much consideration once the symptoms of shortages—those famous long lines at the gas stations—had gone away.

SUMMARY OF RESULTS

The four opinion/news publications analyzed—*The Nation*, *National Review*, *The New Republic*, and the *New York Times Magazine*—focused mainly on government's relationship to the oil industry, measures of competitiveness or indications of excess power or influence, and current problems. The *National Review* was the sole publication to suggest that the energy crisis required less energy problem solving by the government. Also, *The Nation* and *The New Republic* showed special concern about big oil's power aside from any antitrust considerations.

The Nation provided virtually no balance in either argument or source selection, or space allotment. Its predominantly anti-oil industry sources were comprised of about one-half government, one-fourth independent consulting, one-fourth other media, and one-fourth industry sources.

The National Review, in stressing its objections to government intervention into energy matters (except for purposes of protecting national security), used sources predominantly favorable to the oil industry's views. However, this publication divided source space evenly among pro and anti-industry and neutral sources. The *Review* quoted or cited a large proportion of academic or independent oil experts (nearly half its total sources), paying scant attention to either government or oil spokespeople. It also focused on a number of news media sources (about a third of total sources), criticizing them for "demagogy" in presenting oil economics issues.

The New Republic heavily favored anti-industry sources (about two to one) and arguments (nearly eight to one). It favored anti-industry sources in about two-thirds of its articles but balanced source space in the remaining third. In focusing a great deal on government policies and responsibilities, it consulted primarily government sources. Only about 10 percent of its sources were nongovernmental critics or industry figures and an even lower proportion consisted of academic or independent consulting sources (unless one counts the economist who wrote a review for this publication).

The *New York Times Magazine* featured anti-industry and pro-industry arguments in about a four to one ratio; and it utilized almost twice as many anti- as pro-industry sources. But it did balance source space fairly evenly. It relied solely on government (about two-thirds) and industry (about one-third) sources for its information and views. However, one contributor was an independent academic (field unspecified).

REVIEW OF CONTENT ANALYSIS FINDINGS
FOR ALL PERIODICALS

The prognosis for congressional passage of either vertical or horizontal divestiture legislation will probably remain unsettled until well into or beyond the 95th session of Congress in 1977. An important question now is what would be divestiture's effects should it be passed? The resumption of the Federal Trade Commission and a number of senators has been that such legislation would help to restore competition and ultimately provide relief to consumers. The measure's opponents—notably, the oil industry (including both majors and independents), conservative politicians, the Justice Department, and most academic economists who specialize in studying the oil industry's behavioral and structural effects— foresee few additions to competition (with the possible exception of pipelines) and indeed, costly disruptions both to competition and to the production and distribution of oil and its products. Nor do many of the economists and even some FTC lawyers see any benefit in pursuing horizontal divestiture.

From the public's standpoint, who is to be believed? Which sources and arguments have presented the issue most clearly and forcibly? What evidence has most often been brought to bear on the issue? Are the sources of this evidence trustworthy and reliable? Is there any information that has been left out? From this perspective, and in view of the preceding content analysis, it appears that the "facts" presented to the public by the periodical press suffer from some rather serious shortcomings.

The business publications studied provided a relatively balanced but some-how dissatisfying array of numbers of arguments and sources. *Business Week*, for example, following a news format, covered in-the-new figures—the FTC and Senate investigators and well known corporate heads—almost to the exclusion of more moderate and objective sources.* Further, *Business Week* was evidently so conscious of presenting both sides fairly that it actually presented more anti- than pro-industry arguments. *Dun's*, a bit obtuse on oil matters per se, still provided some in-depth looks into broader business-economic issues from a largely industry standpoint. And it did consult some sources who seemed reasonably apolitical or at least not directly affected by the divestiture question. *Forbes*, un-abashedly probusiness and pro-oil-industry, still managed to balance sources—

---

*It should be noted that *Business Week* did eventually consult some of the leading independent academic economists as to their views on divestiture following the March 1973-March 1976 time period of this study. Other publications may have done similar articles, but there was no systematic attempt to find them beyond March 1976. However, divestiture efforts slowed down in the fall and summer of 1976, suggesting that media coverage of the issue did, too.

though certainly not arguments—reasonably well. It, too, included some sources who were knowledgeable about but not directly involved in industry or government actions re divestiture. *Fortune* engaged in some thoughtful analyses of the public ramifications of oil profits and activities; it also employed several independent experts to discuss what had occurred since the embargo; and its arguments and sources were quite well balanced.

One common denominator—and perhaps a flaw—of the business periodicals was their concentration in image. Articles throughout identify corporate image problems as insensitivity, lack of cooperation, and an inability to translate information about business to a hostile Congress or public. Yet the publications themselves largely avoided an analysis of the basis for charges. One exception to this was the profits question, an area of extreme public resentment. However, there are other aspects of the monopoly/competition/divestiture issue that business periodicals, especially, might have been in a position to both understand and elucidate. Perhaps they felt their readers already knew about such things and skipped to the closer-to-home issue of effects. Or their highlighting of Senate and FTC implicit-behavior arguments might have been done to familiarize corporate executives with unknown territory. But why, then, did they not also identify more of the contrary data and sources? If their objective was to give the impression of impartiality and to remain noncontroversial, then they generally succeeded.

The news publications, as one would expect, maintained a relatively high level of objectivity in commenting upon news reports (*Time* having the greatest propensity for editorial asides). But the selection of sources and arguments by the two largest, *Newsweek* and *Time*, suggests that they spent far more time reading the FTC report and listening to senators at news conferences than they did digging behind the scenes—reading hearings transcripts or talking to a few extra sources whose pro-industry biases were less pronounced than those of the typical multinational president. (*Newsweek*'s anti-oil to pro-oil-industry combined source and argument ratio was about two to one; *Time*'s was closer to three to one.) They did, however, go to one kind of source very effectively—the truck drivers, gas station attendants, shoppers, and commuters—to ferret out basic analyses. But what did they, in turn, do to help these sources understand what was happening? To a large extent, the news periodicals mirrored the confusion that created the anger that may have blinded many legislators to cooler, calmer observations. (An exception to this was *Newsweek*'s cool, calm Milton Friedman, the conservative "odd man out.") Perhaps mirroring or reflecting is the nature—or one of the limitations—of news reporting in brief. The loudest squeaking wheels were in Congress and among industry critics, and they received the greatest attention.

The third news periodical, *U.S. News & World Report*—less popular and widely read than the other news weeklies, geared to a more conservative audience, and not too interested in interviewing liberal senators—went to sources whose voices had been lost in the more vocal public clamor. Their views, while supporting industry in many respects, were not as preoccupied with big oil's sins as with longer-term energy issues. In this sense, *U.S. News*, while fairly biased in source selection (but relatively well balanced in argument selection), perceived the news as less unidimensional. Of course, it was also duller than *Time* or *Newsweek*.

The opinion publications were of two types—biased and more biased. While a strong philosophical or political perspective is to be expected, there are still some criteria of fairness and truth seeking by which such publications can be judged. The *Nation* and *National Review* both retain tight editorial control over selection of contributors. Dissenting authors' voices do not appear in either publication. However, in contrast to *The Nation* which does not even discuss contrary views, *National Review* went to the trouble of seeking the bases for various anti-industry charges and asked the views of sources who had little to gain personally from supporting the oil industry's claims of competitiveness and its stand against certain governmental restrictions.

The New Republic, too, at least tried to justify its convictions by showing that the evidence contrary to its opinions was refutable. To do this, it had to present some of those other views. The sense of fairness pervading this publication made some of its conclusions seem less than convincing and left dangling some (though it should be stressed, few) doubts as to which sources were actually correct. Thus, *The New Republic* has probably done its more contemplative readers a service and has probably provided other publications (such as the *National Review*) with some intelligent grist to grind. The *New York Times Magazine* seemed to have had its fancy struck by oil shortages and so received one author's strongly antimajor perceptions with open arms. It also received and displayed (though less prominently) comments from people with dissenting views. And it assigned one of its editors to thoroughly examine the oil industry and describe its day-to-day concerns. As far as it went, and despite the juxtaposition of a fiery versus a modestly designed article, the *Times Magazine* did give readers something to hold on to in trying to understand the oil industry.

In three of the opinion publications (*The Nation*, *National Review* and the *Times Magazine*) there was a tendency to resolve the issue of what or who was right or wrong fairly quickly—let's be done with it. Unless the reader can be assumed to have a definite, preformed opinion (not an altogether unlikely assumption for *The Nation* and *National Review* audiences), such an approach bypasses the reader who might really want to examine both the pro and con facets of an issue. *The New Republic* reached the same conclusions as *The Nation* (the *Times* was, after all, inconclusive), but it did consider more carefully the parameters of the debate on the monopoly/divestiture issue.

In summary, the periodical branch of the media has been conscientious in bringing the energy crisis and allegations against the oil industry to public attention. It has been less conscientious in filling out the details of the divestiture/

monopoly debate and has provided uneven coverage, often lacking in impartial information and perspective. The most visible sources and audible arguments were often seized upon uncritically. Nearly completely ignored was the testimony of experts who had made intensive studies of the oil industry. This omission illustrates the press's uncritical acceptance of the most accessible information. Despite all the talk of inaccessibility of information, this is not entirely the case. Senate subcommittee testimony, for example, is readily available. And as the next chapter illustrates, economic experts seem willing to be contacted for their opinions.

Furthermore, as this content analysis has demonstrated, data can be manipulated reasonably easily to provide whatever perspective the author seeks. The use of "only" or "the astounding figure of" can prejudice whatever percentages or large numbers follow—numbers that in and of themselves, are undoubtedly meaningless to many readers. Thus, data alone will not remedy gaps in explaining an evolving and complex issue like monopoly/divestiture. Finally, with some exceptions, those periodicals that were not satisfied with presenting only the most popular arguments actually did less than they could to balance these popular perceptions by introducing disinterested sources to provide contrary evidence. This tension between fairness and advocacy seems to create a special dilemma for publications whose audiences or editorial staffs are on the defense philosophically or on the outs politically.

There is yet a further unresolved issue: If one could identify those sources and arguments that are likely to be the most objective, how would they compare with what has appeared in the periodical press? To examine this, the next chapter looks at the opinions of academic economics experts and compares them to the views of a sample of print media writers and editors. To what extent do these experts agree with what has been said in the periodical press? Which of these people disagree most? And what do these differences imply for the reporting of economic issues of public importance? These are some of the questions Chapter 7 seeks to resolve.

NOTES

1. "Drain America First," *The Nation*, June 28, 1975, p. 771.

2. "Flying Blind," *The Nation*, January 19, 1974, p. 68.

3. Robert Sherrill, "Invisible Empires: The Multinationals Deploy to Rule," *The Nation*, April 16, 1973, p. 490.

4. Ibid., pp. 493-94.

5. Les Aspin, "The Pumps of Privilege: Big Oil and the Nixon Campaign," *The Nation*, February 16, 1974, pp. 207, 210.

6. Donald Bartlett and James Steele, "The Oil Went Thataway," *The Nation*, March 16, 1974, p. 335.

7. Les Aspin, "The Shortage Scenario: Big Oil's Latest Gimmick," *The Nation*, June 18, 1973, p. 776.

8. Mary Clay Berry, "The Alaskan Pipeline: After Ecology, Monopoly," *The Nation*, November 5, 1973, p. 467.

9. "Flying Blind," op. cit., p. 68.

10. George L. Baker, "Elk Hills Rip-Off: Standard Oil Keeps Popping Up," *The Nation*, April 26, 1975, p. 490.

11. Bennett Harrison, "Two Case Histories: Inflation by Oligopoly," *The Nation*, August 30, 1975, pp. 147-48.

12. "Drain America First," op. cit., p. 772.

13. Sherrill, "Invisible Empires," op. cit., p. 494.

14. Aspin, "The Shortage Scenario," op. cit., p. 776.

15. Berry, "The Alaskan Pipeline," op. cit., p. 468.

16. Louis B. Schwartz, "OPEC and Big Oil: The Malign Collaboration," *The Nation*, February 15, 1975, p. 180.

17. Harrison, "Two Case Histories," op. cit., p. 148.

18. William F. Rickenbacker, "Monopolies and Antitrust," *National Review*, November 23, 1973, p. 1301.

19. Ibid.

20. William F. Buckley, "TV on Energy," *National Review*, February 15, 1974, p. 224.

21. Alan Reynolds, "The Energy Crunch: Ten Answers That Are Obvious, and False," *National Review*, April 26, 1974, p. 479.

22. "More Gas is Less," *National Review*, August 16, 1974, p. 908.

23. "Bleeding the Goose," *National Review*, December 20, 1974, p. 1449.

24. Buckley, "TV on Energy," op. cit., p. 224.

25. William F. Buckley, "A Congressman Objects," *National Review*, March 15, 1974, p. 338.

26. William F. Buckley, "Energy Talk," *National Review*, April 13, 1973, p. 435.

27. "Project Dependence," *National Review*, December 6, 1974, p. 1388.

28. "Pork Barrel Socialism," *National Review*, September 13, 1974, p. 1027.

29. Buckley, "Energy Talk," op. cit., p. 435.

30. Buckley, "TV on Energy," op. cit., p. 224.

31. Eliot Marshall, "Gas Up," *The New Republic*, June 22, 1973, p. 11.

32. Robert W. Dietsch, "What Have the Oil Companies Been Up To?" *The New Republic*, September 28, 1974, p. 16.

33. Melville J. Ulmer, "How to Treat Parasites: Thwarting the Cartel," *The New Republic*, February 15, 1975, p. 10.

34. Josiah Lee Auspitz, "Oil: The Strategic Utility," *The New Republic*, April 26, 1975, p. 14.

35. Melville J. Ulmer, "Captains of Our Fate: The Oil Masters," *The New Republic*, January 5 and 12, 1974, p. 14.

36. Muriel Allen and Richard Levy, "Whose Alaskan Oil?: Arco, Exxon and BP Think It's Theirs," *The New Republic*, July 28 and August 4, 1973, p. 16.

37. "Oil Profits Spree," *The New Republic*, April 27, 1974, p. 5.

38. "Unchain the Beast?" *The New Republic*, March 22, 1975, p. 8.

39. "Out of Control," *The New Republic*, September 20, 1975, p. 9.

40. "The Oil Information Shortage," *The New Republic*, January 26, 1974, p. 7.

41. "Public Oil and Gas," *The New Republic*, February 2, 1974, p. 6.

42. Ulmer, "The Oil Masters," op. cit., pp. 13-14.

43. "Public Oil and Gas," op. cit., p. 6.

44. "Breaking Up Oil," *The New Republic*, December 27, 1975, p. 8.

45. "The Oil Information Shortage," op. cit., p. 7.

46. "Public Oil and Gas," op. cit., p. 6.

47. Robert W. Dietsch, "More Oil Profits: The Industry Outlook," *The New Republic* March 2, 1974, p. 19.

48. M. A. Adelman, review of *Making Democracy Safe for Oil*, by Christopher T. Rand, *The New Republic*, November 15, 1975, p. 27.

49. Brit Hume, "The Case Against Big Oil: Why Was There a Shortage Before the Shortage?" *New York Times Magazine*, December 9, 1973, p. 40.

50. Paul Lewis, "Getting Even," *New York Times Magazine*, December 15, 1974, p. 13.

51. Russell Baker, "The Crisis Crisis," *New York Times Magazine*, January 20, 1974, p. 13.

52. Ibid.

53. Alan Reynolds, "Free-market Oil?" *New York Times Magazine*, January 6, 1974, p. 2.

54. Brit Hume, "Free-market Oil?" *New York Times Magazine*, January 6, 1974, p. 2.

55. Frederick C. Thayer, "Free-market Oil?" *New York Times Magazine*, January 6, 1974, p. 2.

56. Maurice F. Granville, "A Dialogue on Oil," *New York Times Magazine*, May 5, 1974, pp. 94-95.

57. Robert W. Stock, "One Barrel of Oil," *New York Times Magazine*, April 21, 1974, p. 14.

58. Ibid.

7

EXPERT VERSUS MEDIA
PERCEPTIONS OF THE
OIL MONOPOLY QUESTION

How do the perceptions of academic economists who specialize in studying oil industry monopoly issues compare with those of the print media, which select our testimony, facts, and opinions for public presentation? To answer this question and to seek some corroborating evidence (in relation to the findings of the content analysis) about the media's attitudes toward the monopoly/divestiture issue, this portion of the study utilizes a scaled attitude survey.

SAMPLE DESIGN

Before discussing the design of the survey questionnaire it is useful to look at the rationale for selecting the technique and sampling procedures growing out of it. Two primary groups—academic industrial organization economists specializing in oil economics, and journalists who write for or edit newspapers and periodicals—are expected to diverge or differ on whether major, integrated oil companies possess or exercise monopoly control and on whether public perceptions of the issue have been influenced by, or at any rate exposed to, appropriate sources, facts, and arguments.

The first group was composed of those economists who were recognized as knowledgeable authorities on the oil industry and its structure and behavior.*

---

*In fact, a petition signed by economists of varied backgrounds produced opinions largely contrary to the opinions of the oil industry specialists queried in this study. Dr. Walter Adams, Distinguished University Professor, Department of Economics, Michigan

The selection of oil industry specialists does not preclude their being biased. But since they are academics rather than consultants to the oil industry (admittedly some may be or may have been both), they can be viewed as some of the most objective sources one could consult to ascertain the facts, or more importantly, how the facts should be interpreted.

The economists selected for this study were those who presented testimony at either the 1969 or 1975 hearings of Senator Hart's Subcommittee on Antitrust and Monopoly or other congressional hearings; and those additional economists who, although they may not have testified, were listed in the *Directory* of the American Economic Association, under the Industrial Organization heading, and had published or studied extensively in the area of oil economics.* Although this list may not be totally inclusive, it does try to take account of those analysts who were deliberately chosen by Senate investigators to contribute to the public record, and it has recognized, as far as practicable, other analysts who have publically presented academic work on oil monopoly/divestiture subjects.

Using these criteria, 29 economists were chosen and asked to answer a questionnaire on oil monopoly, divestiture, and public perceptions. In the list below, certain symbols have been used to indicate which of the economists have presented testimony. Those with an asterisk are known to have testified before the Hart subcommittee; a plus symbol indicates that they testified before some other congressional subcommittee(s); and a question mark denotes those who are thought (unverified) to have testified before some congressional subcommittee or committee.

Walter Adams *+
Michigan State University

Paul Joskow
Massachusetts Institute of Technology

State University, circulated the petition. It was signed by 155 economists nationwide who supported divestiture legislation. This was read into the *Congressional Record* by Senator Phillip Hart on June 8, 1976. Dr. Adams, along with one other economist included as a questionnaire subject in this study, signed the petition.

*Two economists who testified at the 1969 hearings, Dr. Walter Newton and Dr. Edith Penrose, were excluded from the questionnaire list because they were British economists and because their previous testimony indicated their work might not be germane to a discussion of the U.S. petroleum market per se.

The author wishes to acknowledge the assistance of Professor Richard B. Mancke, director of the energy studies/research program at the Fletcher School of Law and Diplomacy, Tufts University, in identifying, from the American Economic Association listing, those industrial organization economists who—although they had not, or were not known to have, testified at congressional hearings—specialized in oil economics.

M. A. Adelman *+
Massachusetts Institute of Technology

Paul Bradley *
University of British Columbia

Keith Brown ?
Purdue University

Paul Davidson *
Rutgers University

Joel Dirlam *
University of Rhode Island

Thomas D. Duchesneau
University of Maine

Edward Erickson *+
North Caroline State University

H. J. Frank *
University of Arizona

Richard Gordon
Pennsylvania State University

Robert Helms
American Enterprise Institute
(formerly at Loyola College)

Henry Jacoby +
Massachusetts Institute of Technology

Neil Jacoby *+
University of California at
Berkeley

William A. Johnson *+
George Washington University

Alfred E. Kahn *+
New York Public Service Commission (formerly at Cornell)

Wayne Leeman *
University of Missouri

Richard B. Mancke *+
Tufts University

S. L. McDonald +
University of Texas

James McKie +
University of Texas

Walter Mead *+
University of California at
Santa Barbara

A. J. Meyer ?
Harvard University

Edward A. Mitchell *+
University of Michigan

Michael Rieber *
University of Illinois

Thomas Stauffer +
Harvard University

Henry Steele *
University of Houston

Leonard Waverman
University of Toronto

Martin Zimmerman
Massachusetts Institute of Technology

In sum, then, approximately half these economists have testified for the public record. Twelve, or well over a third, have presented testimony opposed to oil industry practices or industry-related government policies. In addition, several others have published works critical of certain aspects of oil industry behavior. Therefore, prior to questionnaire examination, it was not altogether apparent how many of these economists would favor/oppose divestiture or monopoly arguments.

The second group, journalists, was assumed to be more amorphous, with varying degrees of expertise and interest in monopoly/divestiture questions. Thus, this group was selected by the use of sampling techniques. The first step was to choose target publications. Periodicals were thought to hold the most promise for covering the issue in depth and at length. Large-circulation newspapers also seemed probable sources for national issues and would have sufficient staff and space to devote some attention to this particular issue. Thus, target publications included all news and/or opinion periodicals (plus some oil or gas trade publications), as well as all newspapers with a daily circulation of over 300,000 listed in *The Working Press of the Nation*.[1] Periodicals like the *Oil and Gas Journal* and *Energy Management Report* were included along with *Newsweek*, *National Review*, and so on, to gauge the differences between general and specialized publications. Newspapers ranged from the *Wall Street Journal*, *Christian Science Monitor*, and Washington *Post* to the Buffalo *Evening News* and Seattle *Times*—a widespread geographical distribution that encompassed both national and more localized publications from large metropolitan areas. A total of 23 periodicals and 7 newspapers were chosen as target publications.

A more difficult matter was deciding which editorial staff member on each publication should be sent a questionnaire. Since one pragmatic objective of the questionnaire was to produce as large a response as possible, a flexible strategy was used to determine who should respond. Where possible, questionnaires were sent to those editors or writers denoted as energy editor, business/energy editor, or some such title suggesting that this person would be receptive to the questionnaire or familiar with its concepts. Where this was impossible, news editors, managing editors or (if staff or publication was small) editors-in-chief were chosen. Letters to editors included a request that the recipients—if they could not or did not wish to respond—choose a staff member whom they deemed knowledgeable or as the appropriate person to complete the questionnaire.

Thus, the editors were given the opportunity to select the editors or writers who they believed knew the most about the subject or could best reply. A very erudite response, for example, came from the copy desk of a San Francisco newspaper, from a veteran reporter and news manager who ordinarily would not have been selected as "the" person to query. Other editors answered for themselves, stressing that their replies were strictly their own. The hope was to get enough replies—whether they were considered personal, or in keeping with a publication's editorial position, or a combination of both—to see if a pattern of response emerged. Thus, tight control over who was to respond was relinquished in order to encourage as forthright and large a response as possible and to allow the publications to assist in the process of selecting journalists who either had expertise or could effectively present their views.

QUESTIONNAIRE DESIGN

The questionnaire was accompanied by a cover letter and a stamped return envelope. A second mailing was sent to those who failed to respond within a month. The economists' version of the questionnaire (shown here) differed slightly from the journalists' version in that it sought simply the name of the economist, has institution, age, and sex. Journalists were asked, in addition, to cite their current editorial position and the type and length of their journalistic experience. This was done to see if there were any differences among journalists of varying experience or orientation. All subjects were asked to answer the same basic questionnaire and to supply whatever comments they wished.

The questionnaire sought to be fairly specific in determining respondents' degrees of acceptance or rejection of certain statements; thus, it incorporated a 7-point scale to gauge shades of meaning:

Strongly Agree (1)	Agree (2)	Partially Agree (3)	Neutral (4)	Partially Disagree (5)	Disagree (6)	Strongly Disagree (7)

Before discussing the questionnaire rate of return and results, a look at the questionnaire will show what questions were considered important to this study of expert versus media perceptions.

QUESTIONNAIRE ON OIL MONOPOLY/DIVESTITURE

Name _____ (confidential, for identification purposes only)

Institution _____Position _____

Age _____ Sex _____.

☐ Check here if you wish to receive a synopsis of the results of this study.

INSTRUCTIONS: Please check your agreement-disagreement with the following statements using this scale.

Strongly Agree (SA)	Agree (A)	Partially Agree (PA)	Neutral (N)	Partially Disagree (PD)	Disagree (D)	Strongly Disagree (SD)

(continued)

(Questionnaire continued)

1. The major, integrated U.S. oil companies currently possess monopoly power in one or more stages of the oil business.

___(SA)   ___(A)   ___(PA)   ___(N)   ___(PD)   ___(D)   ___(SD)

2. Specifically, the majors exercise monopoly power in . . .

	(SA)	(A)	(PA)	(N)	(PD)	(D)	(SD)
oil production	—	—	—	—	—	—	—
refining	—	—	—	—	—	—	—
marketing of refined products	—	—	—	—	—	—	—
pipelines	—	—	—	—	—	—	—
tankers	—	—	—	—	—	—	—
other energy areas (coal, nuclear, etc.)	—	—	—	—	—	—	—

3. The major, integrated oil companies do not currently possess monopoly power, but they do have the potential for controlling the U.S. petroleum market.

___(SA)   ___(A)   ___(PA)   ___(N)   ___(PD)   ___(D)   ___(SD)

4. The major, integrated oil companies make it difficult for independent, partially integrated companies to compete.

___(SA)   ___(A)   ___(PA)   ___(N)   ___(PD)   ___(D)   ___(SD)

5. The proposal to break up the oil companies' integrated operations, now being made by several senators, is a good idea.

___(SA)   ___(A)   ___(PA)   ___(N)   ___(PD)   ___(D)   ___(SD)

6. These senators are motivated by a desire to promote competition. . .

	(SA)	(A)	(PA)	(N)	(PD)	(D)	(SD)
in all U.S. industries	—	—	—	—	—	—	—
only among oil companies	—	—	—	—	—	—	—

7. A large part of these senators' motivation stems from a desire to be re-elected or elected to a higher office and divestiture is a good campaign issue.

___(SA)   ___(A)   ___(PA)   ___(N)   ___(PD)   ___(D)   ___(SD)

8. The television and radio media have presented fairly and completely the facts necessary to form an opinion as to whether divestiture is desirable.

___(SA)   ___(A)   ___(PA)   ___(N)   ___(PD)   ___(D)   ___(SD)

9. The print media (newspapers, magazines) have presented fairly and completely the facts necessary to form an opinion about divestiture.
   ___(SA)   ___(A)   ___(PA)   ___(N)   ___(PD)   ___(D)   ___(SD)

10. The mass media have provided equal time or space to both the advocates and opponents of divestiture.
    ___(SA)   ___(A)   ___(PA)   ___(N)   ___(PD)   ___(D)   ___(SD)

11. The financial press has been more responsible than the general press in reporting on oil company monopoly and divestiture matters.
    ___(SA)   ___(A)   ___(PA)   ___(N)   ___(PD)   ___(D)   ___(SD)

12. There appears to be a consensus among economic experts as to the efficacy of divestiture.
    ___(SA)   ___(A)   ___(PA)   ___(N)   ___(PD)   ___(D)   ___(SD)

13. These economic experts fail to see the value in divestiture.
    ___(SA)   ___(A)   ___(PA)   ___(N)   ___(PD)   ___(D)   ___(SD)

14. Divestiture of the major, integrated oil companies' various operations would affect the oil industry's performance in these ways. . . .

	(SA)	(A)	(PA)	(N)	(PD)	(D)	(SD)
it would lower prices	___	___	___	___	___	___	___
services would improve	___	___	___	___	___	___	___
companies would be less efficient	___	___	___	___	___	___	___
it wouldn't make much difference	___	___	___	___	___	___	___

15. The oil divestiture issue is sidetracking us from real energy problem solving.
    ___(SA)   ___(A)   ___(PA)   ___(N)   ___(PD)   ___(D)   ___(SD)

Additional Comments: _____

The actual questions—or more precisely, statements—were designed to achieve a number of theoretical and practical goals. One basic theoretical aim was to pose statements that would afford comparison or contrast of respondents' opinion differentiation patterns and their degrees of opinion or attitudinal intensity. A primary practical aim was to determine respondents' policy concerns or issue orientations in specific terms and to elicit the expression of any additional concerns that might lie outside the scope of the questionnaire.

In addition to the general monopoly statement in question 1, question 2 sought respondents' views about degrees of monopoly control in the multiple phases of the oil business. This was done to give respondents an opportunity to say more about the specificity as well as the strength of their convictions. Questions 3 and 4, for example, were designed to discover the "uncertains," those who saw potential for but were not convinced of monopoly control, or those who might not accept monopoly but were chary of the power or bigness of the major, integrated oil companies. Question 5 was designed to distinguish between those who agreed/disagreed with the monopoly argument but nevertheless disagreed/agreed with the divestiture solution.

Questions 6-11 were image questions, seeking essentially intuitive responses to the primary newsmakers in the divestiture debate—senators—and to the media's performance in creating an atmosphere of reason, chaos, or something in between. Question 11 was included to see if the business branch of the press (which has a modicum of expertise in industry matters) was perceived as any different from—or better than—other media forms. Question 12 was aimed at comparing self-perception (for the economists) with the media's perception of how much confusion or conflict exists for economic experts regarding divestiture.

Questions 13-15 were designed to examine people's perceptions of the actual value of divestiture. Would experts find it useful, would the respondents find it useful, or is it sidetracking the nation from the solution of more important energy problems?

RATE OF RETURN OF QUESTIONNAIRES

The returns produced by two mailings (questionnaire, cover letters, and stamped return envelopes both times) were as follows. Mailings were sent to a total of 109 individuals—29 economists and 80 journalists. Sixty-one returned their questionnaires, a response rate of 56 percent. Fifty-two, or 85 percent of these were usable. (Some questionnaires were returned with comments but few or no answers to scaled questions.) Thus, 48 percent of the questionnaires were returned and usable.

Thirty-six of the 80 journalists, or 45 percent, returned their questionnaires. Of these, 31 were usable. Thus, 39 percent of the journalists' questionnaires were returned and usable. Among journalists, 21 (68 percent) of the responses were from newspaper writers or editors and 10 (32 percent) were from periodical writers or editors. So, within publication types, 37 percent of the newspapers and 43 percent of the periodicals queried responded. However, if the trade publications are excluded, periodicals were actually less responsive proportionately than newspapers. Not surprisingly, those closest to the issue (or perhaps those who were less concerned with maintaining objectivity or neutrality) were most willing to reply and tended to answer more parts of the questionnaire.

Twenty-five, or 86 percent, of the economists returned their question-naires or made some sort of reply. Of those returns, 21 were usable. Thus, 72 percent of the economists' questionnaires were returned and usable.

In sum, then, response from economists was far more enthusiastic than from journalists. Newspaper journalists were more responsive than nontrade periodical editors or writers. Many of those who did respond—journalists especially, but economists too—supplied comments or even whole articles or papers to provide explanations beyond the scope of the relatively simple questionnaire.

QUESTIONNAIRE ANALYSIS: TECHNIQUES AND DEMOGRAPHICS

The demographic data showed that of 52 respondents, 50 (economists and journalists) were men and 2 (journalists only) were women. The average ages of the two types of respondents were close—46 for economists and 44 for journalists. The average journalist had more than 16 years of experience. Four of the journalists (13 percent) wrote on business or economic subjects; 7 (23 percent) wrote on energy matters; 2 (6 percent) wrote on environmental or other subjects; one did not specify any area—demographic information other than age and sex; and 17 (57 percent) were general news writers or editors.

As for journalists' current job titles, 16 (53 percent) listed themselves as managing editors or (less frequently) editors-in-chief, 11 (37 percent) were special area editors, and 3 (10 percent) were writers or reporters. Twenty-one (70 percent) represented newspapers and 9 (30 percent) were on periodical staffs.

The statistical tests run on the 15 questions and their subparts were as follows:

1. Cross tabulations of journalists' versus economists' responses to each question and a chi-square test of significance of the differences between the cross tabulations.

2. Cross tabulations of newspaper versus magazine journalists' responses to each question and a chi-square test on cross-tabulation differences.

3. Calculation of the mean scores for journalists and economists on each question using the t-test to show whether the differences between the means were statistically significant at the .05 percent level.

4. Calculation of the means for business/economic/energy journalists versus general news journalists and test of mean differences using the t-test.

5. Calculation of the means for business/economic/energy versus general news journalists plus one environmental and one unspecified writer, again using the t-test to gauge mean differences.

6. Calculation of the means for business/economic/energy journalists versus economists, again using the t-test of significance.

7. Calculation of the means for general news journalists versus economists, again using the t-test.

ECONOMISTS' VERSUS JOURNALISTS' RESPONSES

Percentage Score Comparisons

The data on questions 1-15 led to a number of interesting comparisons and contrasts. Before proceeding with statistically significant differences, a few examples of the percentage scores of different groups and subgroups on several questions are presented below. The most basic difference occurred between economists and journalists concerning the essential nature of the petroleum industry and how it should be treated. Among economists, 76 percent ranged from partial to strong disagreement with the charge that the majors possessed monopoly power. Forty-three percent voiced strong disagreement with this statement. The journalists' responses were much more spreadout, with 54 percent strongly to partially agreeing with the monopoly charge (21 percent agreed strongly), 4 percent neutral, and 43 percent ranging from partial to strong disagreement (18 percent disagreed strongly).

Ratings of statements 2 through 7 show that journalists were not as discriminating as economists in distinguishing the degree of monopoly power that might exist in different stages of the oil business. At least half the journalists thought the major oil companies exercise monopoly power in oil refining (57 percent), marketing of refined products (54 percent), and pipeline operations (50 percent). Less than half the journalists thought that the majors had monopolized crude oil production (43 percent), tankers (43 percent), and other energy areas such as coal and nuclear (36 percent).

A sizeable majority of the economists failed to perceive monopoly problems in any state of the oil business. Specifically, 33 percent agreed that the majors exercise some monopoly power in pipelines; 24 percent held a similar view for crude oil production and marketing of refined products; 19 percent saw monopoly problems in the refining stage; and only 10 percent (two economists) thought there were any anticompetitive effects in the tanker market or in other energy areas.

There was also a marked difference between economists and journalists on the extent to which major oil companies had inhibited independent, partially-integrated competitors. While 67 percent of the economists disagreed with this statement (33 percent strongly), 68 percent of the journalists (33 percent of the economists) strongly endorsed the notion that majors make it difficult for independents to compete.

Ninety-one percent of the economists tended to disagree with statement 9, "The print media (newspapers, magazines) have presented fairly and completely the facts necessary to form an opinion about divestiture." Sixty-three percent of the journalists concurred with this assessment. However, both economists and journalists (80 percent of both groups) tended to agree with statement 10, which held that the financial press had been more responsible than the general press in reporting the oil monopoly/divestiture issue.

Responses to statements 11, 12, and 13 suggest that economists may think they disagree with one another on the issue but are actually in greater concert than either they, or especially journalists, think. Almost half (48 percent) of the economists disagreed that consensus existed among economic experts as to the efficacy of divestiture. Sixty-six percent of the journalists (13 percent strongly) also perceived conflict among the experts. However, 58 percent of the economists (21 percent strongly) agreed that "These economic experts fail to see the value in divestiture," whereas only 32 percent of the journalists thought that this was the case.

Economists were also more in consensus and more convinced than journalists that divestiture would do little to lower prices or improve services. Ninety percent of the economists (55 percent strongly) disagreed that divestiture would lower prices compared to 63 percent (27 percent) of the journalists. Although both groups were pessimistic about divestiture's effects, economists were much more resolute that it would not improve petroleum market conditions.

Finally, responses to statement 15 and 16 show the stronger objections economists have to the divestiture proposal compared to journalists. Eighty-six percent of the economists (57 percent strongly) disagreed that the proposal was a good idea. Ninety-five percent (all but one economist, who partially disagreed) agreed with the statement that the divestiture issue is sidetracking the United States from solving its real energy problems. A less assured 48 percent of the journalists questioned the wisdom of the proposal and 74 percent agreed (36 percent strongly) that divestiture is interfering with energy problem solving.

The next section analyzes the statistically significant differences that crop up not only between economists and journalists, but among journalists, and between economists and certain types of journalists. The following material, in tabular form, illustrates the frequency and degree of differences.

The tables show the groups being compared, the number of cases (responses to each question), the groups' mean scores, the standard error (or standard deviation) of the mean scores, and the significance levels of the differences between the mean scores of each group. For example, a significance level of .01 would indicate that the probability is only 1 in 100 that both groups would have the same mean score on a given question. In short, the lower the significance level, the less likely the two groups are to concur. Significance levels of .05 (5 percent) customarily are considered to be statistically significant. In the following tables all significance levels below .05 (5 percent) are considered to be statistically significant. In the following tables all significance levels below .05 (5 percent) are considered to be statistically significant. Significance levels below .05 are denoted by a plus (+) symbol.

Statistically Significant Differences

Tables 7.2-7.3 show that there are a number of statistically significant differences between economists and journalists (especially general news journalists) on oil monopoly/divestiture issues. On the most basic question—whether the oil industry is monopolized—economists disagreed that the majors are monopolists whereas the journalists tended toward very mild agreement or neutrality (a result of a split on the issue). The difference between the means of the two groups is significant at the .01 level.

On questions as to whether monopoly exists in specific oil operations, the differences were also statistically significant in the majority of cases. In all cases—production, refining, marketing, tankers, pipelines, and other energy areas—the economists perceived fewer monopoly problems than did the journalists.

Table 7.2 shows some similar though less pronounced differences between different types of journalists: general news writers versus special area writers on business, economic, or energy subjects. The special area writers were much closer to economists in their views of monopoly/divestiture. And as Table 7.3 shows, the economists (who, except for two individuals, were frequently in consensus) were most at variance with the general news writers. That is, economists expressed few strong opinions that monopoly exists, whereas those news journalists who accepted the notion at all, tended to accept it rather strongly.

Some exceptions to the general pattern appear in the tables. In question 3, for example, both business and general news journalists tended to disagree with the statement that "the major, integrated oil companies do not currently possess monopoly power but they do have the potential for controlling the U.S. petroleum market." This result seems to be due to ambiguity of the question itself. Judging from their other responses, it probably means that general news journalists were less willing to accept the idea that there is no present monopoly while business journalists did not wish to attribute potential monopoly power to the majors.

Question 12 produced another unusual response. Not surprisingly, economists perceived greater consensus among economic experts than did all journalists. But general news journalists perceived greater consensus than did business journalists. To speculate, this result may be due to business journalists being more in touch with the issue and thus perceiving the same conflicts that oil industry members perceive being generated by economic experts from the Federal Trade Commission, Senate subcommittee staffs, and so on, rather than from the

TABLE 7.1

Comparisons of Answers to Monopoly/Divestiture Questionnaire: All Economists versus All Journalists

Group	Number of Cases	Mean Score	Standard Error of the Mean	Significance Level of Mean Differences	
1. The major, integrated U.S. oil companies currently possess monopoly power in one or more stages of the oil business.					
Economists	21	5.52	.418	.010	+
Journalists	28	3.89	.437		
2A. The majors exercise monopoly power in oil production.					
Economists	21	5.67	.449	.063	
Journalists	28	4.43	.470		
2B. The majors exercise monopoly power in refining.					
Economists	21	5.71	.379	.002	+
Journalists	28	3.79	.431		
2C. The majors exercise monopoly power in marketing of refined products.					
Economists	21	5.62	.399	.007	+
Journalists	28	3.89	.467		
2D. The majors exercise monopoly power in pipelines.					
Economists	21	4.67	.422	.213	
Journalists	28	3.93	.405		
2E. The majors exercise monopoly power in tankers.					
Economists	20	6.40	.311	.000	+
Journalists	28	4.32	.415		
2F. The majors exercise monopoly power in other energy areas.					
Economists	21	5.95	.334	.022	+
Journalists	28	4.71	.401		
3. The major, integrated oil companies do not currently possess monopoly power but they do have the potential for controlling the U.S. petroleum market.					
Economists	19	5.95	.329	.032	+
Journalists	28	4.89	.342		
4. The major, integrated oil companies make it difficult for independent, partially integrated companies to compete.					
Economists	21	5.05	.439	.003	+
Journalists	31	3.32	.342		

(continued)

(Table 7.1 continued)

Group	Number of Cases	Mean Score	Standard Error of the Mean	Significance Level of Mean Differences

5. The proposal to break up the oil companies' integrated operations, now being made by several senators, is a good idea.

Economists	21	5.95	.375	.022 +
Journalists	31	4.61	.427	

6A. These senators are motivated by a desire to promote competition in all industries.

Economists	20	5.40	.432	.358
Journalists	23	4.87	.374	

6B. These senators are motivated by a desire to promote competition only among oil companies.

Economists	21	4.67	.504	237
Journalists	26	3.92	.359	

7. A large part of these senators' motivation stems from a desire to be reelected or elected to a higher office, and divestiture is a good campaign issue.

Economists	21	1.90	.217	.034 +
Journalists	31	2.65	.260	

8. The television and radio media have presented fairly and completely the facts necessary to form an opinion as to whether divestiture is desirable.

Economists	20	6.15	.182	.695
Journalists	29	6.03	.230	

9. The print media (newspapers, magazines) have presented fairly and completely the facts necessary to form an opinion as to whether divestiture is desirable.

Economists	21	5.67	.187	.041 +
Journalists	30	4.87	.331	

10. The mass media as a whole has provided equal time or space to both the advocates and opponents of divestiture.

Economists	20	5.60	.328	.403
Journalists	29	5.21	.331	

11. The financial press has been more responsible than the general press in reporting on oil company monopoly and divestiture matters.

Economists	20	2.50	.235	.404
Journalists	29	2.83	.310	

Group	Number of Cases	Mean Score	Standard Error of the Mean	Significance Level of Mean Differences	

12. There appears to be consensus among economic experts as to the efficacy of divestiture.

Group	Number of Cases	Mean Score	Standard Error of the Mean	Significance Level of Mean Differences	
Economists	21	3.81	.450	.011	+
Journalists	29	5.28	.317		

13. These economic experts fail to see the value in divestiture.

Group	Number of Cases	Mean Score	Standard Error of the Mean	Significance Level of Mean Differences
Economists	19	3.11	.382	.220
Journalists	25	3.68	.256	

14A. Divestiture of the major, integrated oil companies' various operations would lower prices.

Group	Number of Cases	Mean Score	Standard Error of the Mean	Significance Level of Mean Differences	
Economists	20	6.05	.336	.027	+
Journalists	30	4.90	.376		

14B. Divestiture would result in improved services.

Group	Number of Cases	Mean Score	Standard Error of the Mean	Significance Level of Mean Differences
Economists	20	5.85	.319	.123
Journalists	31	5.13	.330	

14C. Divestiture would make companies less efficient.

Group	Number of Cases	Mean Score	Standard Error of the Mean	Significance Level of Mean Differences
Economists	20	2.90	.447	.460
Journalists	31	3.32	.348	

14D. Divestiture wouldn't make much difference.

Group	Number of Cases	Mean Score	Standard Error of the Mean	Significance Level of Mean Differences
Economists	21	4.57	.440	.288
Journalists	30	5.17	.336	

15. The oil divestiture issue is sidetracking us from real energy problem solving.

Group	Number of Cases	Mean Score	Standard Error of the Mean	Significance Level of Mean Differences	
Economists	21	1.76	.238	.020	+
Journalists	31	2.81	.363		

Source: Compiled by the author.

TABLE 7.2

Comparisons of Answers to Monopoly/Divestiture Questionnaire: General News-Environment versus Business-Energy Journalists

Group	Number of Cases	Mean Score	Standard Error of the Mean	Significance Level of Mean Differences	
1. The majors possess monopoly power.					
General News	16	3.25	.520	.041	+
Business	11	5.09	.667		
2A. The majors have monopoly power in oil production.					
General News	17	3.88	.568	.061	
Business	10	5.70	.716		
2B. The majors have monopoly power in refining.					
General News	17	2.94	.441	.005	+
Business	10	5.50	.654		
2C. The majors have monopoly power in marketing.					
General News	17	2.88	.469	.002	+
Business	10	5.90	.657		
2D. The majors have monopoly power in pipelines.					
General News	17	3.53	.463	.123	
Business	10	4.90	.706		
2E. The majors have monopoly power in tankers.					
General News	17	3.82	.502	.045	+
Business	10	5.50	.601		
2F. The majors have monopoly power in other energy areas.					
General News	17	4.12	.453	.011	+
Business	10	6.10	.547		
3. The majors have no present monopoly but they do have the potential.					
General News	16	4.38	.455	.121	
Business	11	5.45	.493		
4. The majors make independent competition difficult.					
General News	19	2.74	.314	.024	+
Business	11	4.55	.652		
5. The proposal to divest is a good idea.					
General News	19	4.05	.543	.022	+
Business	11	5.91	.530		
6A. Senators desire competition in all industries.					
General News	14	4.36	.427	.077	
Business	8	5.88	.666		

	Group	Number of Cases	Mean Score	Standard Error of the Mean	Significance Level of Mean Differences	
6B.	Senators desire competition only in oil industry.					
	General News	17	3.65	.402	.362	
	Business	8	4.50	.802		
7.	Senators want to be elected; divestiture is a good campaign issue.					
	General News	19	3.05	.363	.008	+
	Business	11	1.82	.226		
8.	TV and radio have presented the facts.					
	General News	17	5.71	.361	.071	
	Business	11	6.45	.157		
9.	The print media have presented the facts.					
	General News	18	4.28	.403	.049	+
	Business	11	5.64	.509		
10.	The mass media have provided equal time, space to divestiture advocates/ proponents.					
	General News	17	5.12	.410	.931	
	Business	11	5.18	.600		
11.	The financial press has been more responsible re divestiture.					
	General News	17	3.00	.420	.111	
	Business	11	2.18	.263		
12.	Consensus exists among experts.					
	General News	17	4.82	.422	.048	+
	Business	11	6.09	.436		
13.	Experts fail to see the value in divestiture.					
	General News	16	3.81	.319	.472	
	Business	8	3.38	.498		
14A.	Divestiture would lower prices.					
	General News	19	4.42	.435	.013	+
	Business	10	6.20	.490		
14B.	Divestiture would make services improve.					
	General News	19	4.58	.385	.001	+
	Business	11	6.45	.282		
14C.	Divestiture would make companies less efficient.					
	General News	19	3.58	.407	.353	
	Business	11	2.82	.685		
14D.	Divestiture wouldn't make much difference.					
	General News	19	5.05	.378	.858	
	Business	10	5.20	.712		
15.	Divestiture is sidetracking us from energy problem solving.					
	General News	19	3.00	.439	.208	
	Business	11	2.09	.547		

Source: Compiled by the author.

TABLE 7.3

Comparisons of Answers to Monopoly/Divestiture Questionnaire: General News Journalists versus Economists

Group	Number of Cases	Mean Score	Standard Error of the Mean	Significance Level of Mean Differences	
1. The majors possess monopoly power.					
General News	14	3.43	.571	.006	+
Economists	21	5.52	.418		
2A. The majors have monopoly power in oil production.					
General News	15	4.13	.608	.052	
Economists	21	5.67	.449		
2B. The majors have monopoly power in refining.					
General News	15	3.07	.483	.000	+
Economists	21	5.71	.379		
2C. The majors have monopoly power in marketing.					
General News	15	3.00	.516	.000	+
Economists	21	5.62	.399		
2D. The majors have monopoly power in pipelines.					
General News	15	3.60	.524	.123	
Economists	21	4.67	.422		
2E. The majors have monopoly power in tankers.					
General News	15	3.73	.547	.000	+
Economists	20	6.40	.311		
2F. The majors have monopoly power in other energy areas.					
General News	15	4.27	502	.010	+
Economists	21	5.95	.334		
3. The majors have no present monopoly but they do have the potential.					
General News	14	4.21	.505	.009	+
Economists	19	5.95	.329		
4. The majors make independent competition difficult.					
General News	17	2.76	.349	.000	+
Economists	21	5.05	.439		
5. The proposal to divest is a good idea.					
General News	17	4.12	.606	.016	+
Economists	21	5.95	.375		
6A. Senators desire competition in all industries.					
General News	12	4.50	.485	.177	
Economists	20	5.40	.432		

Group	Number of Cases	Mean Score	Standard Error of the Mean	Significance Level of Mean Differences	
6B. Senators desire competition only in the oil industry.					
General News	15	3.67	.454	.150	
Economists	21	4.67	.504		
7. Senators want to be elected; divestiture is a good campaign issue.					
General News	17	2.82	.356	.036	+
Economists	21	1.90	.217		
8. TV and radio have presented the facts.					
General News	15	5.53	.389	.166	
Economists	20	6.15	.182		
9. The print media have presented the facts.					
General News	16	4.13	.437	.004	+
Economists	21	5.67	.187		
10. The mass media have provided equal time, space to divestiture's advocates/ opponents.					
General News	15	4.93	.441	.235	
Economists	20	5.60	.328		
11. The financial press has been more responsible.					
General News	15	3.07	.473	.295	
Economists	20	2.50	235		
12. Consensus exists among experts.					
General News	15	4.80	.470	.138	
Economists	21	3.81	.450		
13. Experts fail to see the value in divestiture.					
General News	14	3.79	.366	.208	
Economists	19	3.11	.382		
14A. Divestiture would lower prices.					
General News	17	4.53	.478	.014	+
Economists	20	6.05	.336		
14B. Divestiture would make services improve.					
General News	17	4.59	.429	.025	+
Economists	20	5.85	.319		
14C. Divestiture would make companies less efficient.					
General News	17	3.53	.447	.326	
Economists	20	2.90	.447		
14D. Divestiture wouldn't make much difference.					
General News	17	5.18	.413	.323	
Economists	21	4.57	.440		
15. Divestiture is sidetracking us from energy problem solving.					
General News	17	3.00	.462	.025	+
Economists	21	1.76	.238		

Source: Compiled by the author.

academic economists who have been identified in this study as most knowledg-able.

The two types of journalists also were closer to one another in their re-sponses to question 15. Journalists—all types—were more inclined than econo-mists to see the divestiture debate as an important, central energy issue, and not merely as interference with energy problem solving. This is probably a natural outcome, for divestiture and monopoly charges have provided the press with many a story. Furthermore, outside the narrow confines of academic thought, greater attention is understandably paid to the obvious—high profits, high prices, and a long history of influence and power—than to strict analysis of what is or is not monopoly power and how it can be neutralized. And finally, economists' solutions tend to be intricate and slow. It is not surprising that the press showed an affinity for a decisive issue, whether they agreed with the solution or not.

There also are several questions on which, notably, economists and jour-nalists agreed with one another. For example, both groups rated television and radio as the worst suppliers of information about divestiture/monopoly and gave the financial press the most credit. However, journalists thought more highly of the skills of the business press than did economists. They all agreed (slightly) that the economic experts saw little value in divestiture. Again, however, the economists as a group tended to agree with this statement, whereas the journal-ists were more split, with about a third agreeing, half neutral or unsure, and the rest disagreeing.

The journalists and economists had mixed responses on the effects divesti-ture would have (questions 14A-14D). Journalists (and news journalists) were more confident than economists (and business journalists) that divestiture would lower prices. News journalists, especially, held greater (though faint) hope for improved services than either economists or business journalists, the latter were the most pessimistic about services improving following divestiture. All groups tended to agree that divestiture would make companies less efficient (with busi-ness journalists most positive this would occur, followed by economists). And they all tended to disagree or partially disagree with the statement, "divestiture wouldn't make much difference." Judging from previous statement ratings this can be interpreted to mean that economists and business journalists predicted that divestiture would have chaotic and unfavorable consequences, while general news journalists foresaw its creating a somewhat more competitive oil industry.

Journalists' and Economists' Comments

A look at some of the economists' and journalists' open-ended responses should help elaborate upon how they viewed the totality of the issue. As the journalists' comments in particular show, acceptance (or suspicion) of monopoly power may exist, but how the nation should resolve such a problem mystifies some of them. As several journalists suggested, "I only know what I read in the papers," and this was thought inadequate for coming to a conclusion. Also, there was a sense of reserve, a hesitancy to commit to an opinion. One energy journalist admitted he was reluctant to do so for fear of compromising his objectivity. But most who supplied comments knew what they believed. As the one environmental writer for a newspaper wrote,

> The large integrated companies now control the flow of petroleum products from the wellhead through the refining process to the point of sale (service stations). This gives them a competitive advantage that is denied to small, independent operators. However, I think the press as a whole lacks the experience and expertise necessary to spell out and interpret the complex interrelationships involved. This is particularly true of television and radio, though newspapers and magazines have not distinguished themselves by their coverage either. Because of this, the public does not really understand what is involved, and as long as that is true politicians will avoid it as a campaign issue. Also, divestiture does not go to the heart of the energy problem; while it might involve incentives to *produce* petroleum products, it will not have that much effect on the *consumption* of those products which—in my opinion—is the real problem.

But an economist perceived a different real problem:

> The real energy problem is a failure of public policy to allow the market mechanism to work. We do not have an energy crisis. We have a policy crisis. If we eliminated all present regulation and differential tax treatment, the "energy crisis" would be over in six months. Divestiture is silly. All responsible economists would agree that the oil and gas industry is effectively competitive.

A Newspaper (managing/news) editor saw the oil industry as only one of many offenders.

> The oil industry is, like much of American industry, an oligopoly, which has the potential to control production, prices, etc. However, strong federal regulartory checks *can* prevent price-collusion, plotted production or restrictions and the gamut of illegal practices—if government is alert. A coming problem—and coming soon—is the oligopoly power concentrated in the food industry. . . . Economists perhaps—and not journalists—can provide better answers to your questions.

But another newspaper journalist had an opposite view:

My views are based on the general proposition that it's impossible to defend the idea that a monopoly exists where nine or more companies are competing. The idea of vertical integration is not in itself counter-competitive: e.g., when consumerists criticize food prices they are likely to take the opposite tack and say food prices are high because the wicked "middleman" intervenes between the virtuous farmer and the noble consumer. The other day it was revealed that for the first time in history the U.S. imported more oil than it produced: it didn't immediately seem to occur to any one that the reason might be the failure to take the ceilings off domestic "old" oil.

A newspaper editorial writer tended to agree with the monopoly charge but had reservations about divestiture.

The questions you pose are difficult policy matters. I am a student of economics (with an M.A. in the field) and I have a visceral feeling that the oil companies have oligopoly power if not monopoly power. Proving that under the antitrust laws is another matter, however, and we have too little research and data. To simply pass a law breaking up the oil industry seems to me an exercise in legislative vigilantism.

Another editorial page editor, wary of the potential for monopoly control but strongly opposed to divestiture, also expressed some doubt about whether the facts can be ascertained or comprehended. He wrote:

Have you influenced your questionnaire by starting from the premise that a monopoly exists? At any rate, this is a highly complex subject with such ramifications that I doubt whether the public or we in the media fully understand it. I have talked to Sen. Gary Hart [a divestiture supporter] about the subject and remain unconvinced of the need for divestiture.

A newspaper news editor, although agreeing with the monopoly charge and divestiture, saw little hope for its succeeding:

The daily press has given less than full coverage of divestiture, either for or against and few newspapers to my knowledge have taken my editorial stands of strength eitherpro or con (of course, I'm not familiar with the oil-producing area newspapers). I feel that the bulk of the American consuming public has a poor opinion of any oil company, largely engendered by the handling of oil shortages, with its attendant high profits of two years ago—however, with the power of lobbyists, etc. it is doubtful, to me, that divestiture has much of a chance in Congress, now or in the future.

An economist, also touching on the question of public lack of understanding (which many respondents seemed to concur on even though they differed in their personal attitudes), contended that shortages have not been caused
by the domestic industry. He commented:

> . . . The real issue is increasing domestic supply. The world export
> market is now controlled by a strong cartel of producing countries
> whose price policy can only be construed as economic warfare.
> (Price is a hundred times production cost, and even higher than a
> long run monopoly price). Increasing import dependence on our part
> is a factor which greatly increases the chances for a devastating oil
> war. We need to develop synthetics and other alternative energy
> sources, as well as to devise policies to increase domestic oil and gas
> discoveries and production. The absolute perfecting of competition
> within the oil industry, by itself, would do little to increase domestic
> supply. This fact is not widely understood by the public, but is
> understood by some of the backers of divestiture bills. . . .

An editor of an independent (producing) trade publication argued similarly:
"Divestiture would not benefit the consumer, the independent oil people, or
the overall effort to secure domestic supplies of energy. It would help destroy
the ability of the oil and gas industry to supply energy."

An economist who partially agreed with the monopoly argument nevertheless opposed blanket divestiture. He wrote:

> Combining all aspects of divestiture confuses the issue and makes
> judgment difficult. I would favor separation of retailing, and pos-
> sibly pipelines, but not production, refining and wholesaling.
>
> I also oppose divesting coal from oil and gas ownership but
> favor strict application of the Sherman and Clayton Acts to both.
>
> The critical role of government policies in *supporting* monop-
> oly is completely ignored in the questionnaire.

Another economist, who viewed marketing and pipelines as somewhat anti-competitive, nevertheless perceived an improvement at the retail level.

> Retail margins might be somewhat lower if there were fewer but
> larger service stations versus the present small stations.
>
> Current conditions seem more competitive at retail than they
> have been for a long time. There is a spread of 6–8 cents/gallon be-
> tween self-service independent stations and full service majors.

An economist who had some agreement with the monopoly charge and
agreed with the divestiture solution did so because he regarded it as a fundamental solution, in spite of complicating factors. He wrote:

It is impossible to predict the consequences of divestiture. It might lower crude prices, or prices at the pump, but there are so many variables—Arab marketing policies, U.S. price controls, etc.—that one simply has to assume that more vigorous competition will be preferable to oligopoly planning, in the long run.

A financial writer (unlike most of the other financial writers in that he supported monopoly arguments) opposed divestiture because:

The probable result of divestiture would be the freeing of billion dollar companies in each segment from anti-trust restraints. For instructive guidance see the number of "independents" who are testifying *for* IBM in the current antitrust suit.

Finally, an economist and a journalist (from a conservative periodical) had similar statement scores and expressed a similar thought. According to the journalist:

The issue of divestiture is a form of scapegoating to draw attention away from Congress' own responsibility for most of our energy problems (i.e., price controls and environmental policies which increased demand for oil and natural gas while discouraging production of domestic supplies).

Said the economist, "Divestiture is Congress' effort to find a scapegoat for its own ineptitude."

NOTE

1. Milton Paule, ed., *Working Press of the Nation*, 27th ed. (Burlington, Iowa: The National Research Bureau, 1976).

8

THE OIL MONOPOLY/ DIVESTITURE ISSUE: IMPLICATIONS AND RECOMMENDATIONS

No lesson seems to be so deeply inculcated by the experiences of life as that you never should trust in experts. If you believe the doctors nothing is wholesome; if you believe the theologians nothing is innocent; if you believe the soldiers nothing is safe. They all require to have their strong wine diluted by a very large admixture of insipid common sense.

> Lord Salisbury in a letter to
> Lord Lytton, Viceroy of India

GENERAL FINDINGS

The results of this study of the oil monopoly/divestiture issue emphasize the disparity that exists between a large majority of the academic economists who specialize in studying the oil industry and those who interpret the actions of the oil industry to the general public. This suggests, in turn, that the information periodical readers (and those people with whom they communicate or whom they influence) have been receiving is incomplete, often lacking in nuance or more than one perspective, and possibly even distorted or misrepresented. All three portions of the study—the background discussion, the content analysis, and the questionnaire results—point to a central conclusion: the press experiences difficulty in, or possibly even resists, identifying objective, primary sources. To some extent the periodical press also has served to amplify the views of only a limited repertoire of individuals and to focus disproportionate attention on conflict and confusion.

The reader of periodical literature, who is exposed to a limited or one-sided set of sources—particularly industry critics, who have become permanent fixtures in much reporting of business matters—may begin to find validity in (or reinforce existing suspicions of) charges of monopoly or oligopoly, squeezing, withholding of oil and gas supplies, conspiracy or collusion, and so on. Unless they were true, why would such charges be heard over and over again? One simple answer to the question, suggested by this study, is that such charges are heard repeatedly because the press presents them repeatedly. The periodical press presents them often because it has relied heavily on traditional newsmakers—most notably politicians—to create and interpret the issue for the public and the press.

Political figures may be well-informed sources, steeped in the subjects they investigate and bring to public attention. But because they are elected officials, they must rely not only on evidence but also upon party philosophy and reelection considerations. The purpose of the Senate antitrust and other subcommittees is to initiate and pass antitrust legislation and to assess how well existing legislation has been applied. Similarly, the Federal Trade Commission, although a fact-gathering body containing people with industrial economics backgrounds, is also charged with bringing antitrust suits. In contrast, the academic economist's concern, generally, is first to analyze what happens and, if a market problem is observed, only then to focus on whether an industry's modus operandi should be altered to bring about a better economic outlook. If one looks at all the available evidence relating to oil company monopoly charges—much of which congressional committees have examined—frequent repetitions of the monopoly charge appear to be inappropriate. Thus, the decision of a number of politicians to press for divestiture legislation seems ill founded.

This study provides documentation that portions of the periodical press have ignored much of the available evidence relevant to evaluating oil monopoly charges. Rather, they have accepted many of the divestiture advocates' statements at face value, highlighted the verbal battles between congressional committee leaders and industry spokesmen or between government officials of opposing political parties, and to no small extent, accenturated the dissent and feeling of public unease the issue was capable of stirring. Now conflict in news stories is much of what news is all about (conflict comprised and gave impetus to much of this study). But the point this study has hoped to illustrate is that the kinds of factual conflict or dissonance that surround the issue have not been properly identified by the press.

Obviously, the public cannot be expected to know what is happening, and convey its opinions to national leaders, when discord and dissonance become the standard for news coverage. Nor can the readers of opinion publications arrive at some sort of consensus on the factual bases of particular issues when certain positions are classified, often unjustifiably, as either liberal or conservative. It is at this juncture that impartial, unemotional, and factually comprehensive input would be helpful. It would help not only to clarify an issue's facts and ramifications—many of which would not be evident to the average person—but to inject calm into a cacophonous debate. Furthermore, expert, relatively unbiased views can take some of the burden of explanation off the shoulders of newspeople who are often generalists assigned to gain expertise in a field.

Unfortunately, it is this element of expertise that so often has been absent in many of the periodicals' discussions of oil industry structure and behavior. Expertise has been absent, but not conspicuously so, because the press has not made it apparent that all considerations and sources have not been taken into account. Limiting the public's access to imparital, knowledgeable sources and arguments ultimately confines or channels public understanding and opinion-forming processes. Admittedly, those economic experts who have made intensive studies of the oil industry are not the only or even the primary sources one would want to contact for information on a political-economic issue like oil divestiture. But government leaders and spokespersons for public-interest groups, likewise, are not the only sources, and certainly not the most impartial. To use a photographic analogy, the academic economists who specialize in observing the oil industry are still stored on negatives while the officials and politicians who are involved in a multiplicity of projects appear on 8 x 10 glossies.

The content analysis portion of this study also has shown that over a fairly extensive time period, some periodicals persisted in presenting similar arguments, recycling themes, and even the same headlines (the most memorable being "Is the Shortage Just Gas?"), and foundering in the same morass of confusion, accusation, or defensiveness. Yet the oil industry has been in considerable upheaval and change due to a variety of factors: the growing costs of producing domestic gas and oil; unabated increases in demand due in large part to inadequate and poorly executed conservation measures; reliance on costly oil imports; an inability to dislodge OPEC from its pricing system; the reduction in special industry privileges, much of which occurred around the time of the OAPEC embargo; much stricter federal regulation of oil companies' pricing; continued antitrust actions or investigations; environmental constraints on prospective energy producers; harsh criticism and prosecution of oil companies involved in campaign slush fund operations; and social inhibitions on corporate demeanor and activities.

Many of the strictures or punitive actions just alluded to may have been necessary for the public good. But deservedness is not the real issue; rather it is that the industry has been evolving and had rectified many of the abuses it was still being accused of by some portions of the periodical media examined in this study. These industry changes, as well as ongoing practices, and the effects of both, are the things that the economists questioned in this study have been chronicling. Their conclusions as to what arguments are appropriate are far more consensual and resolute than the periodical press would lead us to believe. The questionnaire's results confirm that the great majority of these economists agree that oil company monopoly control, with the possible exception of pipeline ownership, is not a great problem and that this issue is one that should be dispensed with so that more difficult, and in their estimate, much more important energy problems can be approached. Few of these economists would disagree with the assertion that the prolongation of the oil monopoly/divestiture issue has been detrimental to public understanding of the larger energy picture, has dulled public sensibilities to thinking about energy policies in constructive ways, and may even have interfered with public cooperation in meeting energy problems.

Blanket condemnation of the efforts and achievements of the periodical press in covering oil economics is, obviously, inappropriate. Several of the publications analyzed—most notably *Fortune*, the *National Review*, and *The New Republic*, plus portions of *Business Week*, *Forbes*, *Newsweek*, *U.S. News & World Report*, and the *New York Times Magazine*—went to varying degrees of trouble to depict the intricate mosaic of oil industry structure and behavior, public concerns, and energy policy making. The first three publications, especially, included reasonable amounts of the kinds of information or interpretations that oil industry economic specialists would emphasize. (However, *The New Republic* generally rebutted such arguments and presented many arguments presumptive of anticompetitiveness.) A number of the publications studied also tried to balance presentations in terms of sources, arguments, or both.

The problem that arises, however, is that numerical balance of viewpoints is not always sufficient. It is insufficient when spatial balance of sources is askew. But it is also insufficient in a subtler way: namely, that profit-motivated oil industry sources, often used to provide a does of the "other side," really do not have the air of credibility that is achieved by "disinterested," sources such as elected government officials, nonelected government prosecutors and investigators, or industry critics of the environmental or consumer protection variety. (It should be noted, further, that these are groups that, although they can be viewed as less-vested interests, are by no means without political clout.) For those business publications presenting industry's views through what they regarded as knowledgeable and respected leaders, balance and insight could be attained. But for the general reader whose views of corporate skullduggery have already become well ingrained (partially as a result of media exposure?), a corporate source is often viewed as unreliable. Further, although a corporate source may have a strong economics background (if so, this was seldom mentioned) and be quite in touch with the pertinent facts, his version of an economic situation is most likely viewed as distorted or possibly even phony.

If many of the findings of this study can be regarded as valid evidence of the nature of periodical press coverage of the oil monopoly question and divestiture-related efforts, a number of implications arise for public policy makers who often provide the substance upon which news is based, and for journalists who convey and interpret policy matters to the reading public. The two groups—policy makers and journalists—are often intertwined in providing the perspective on issues that filter down to the public.

For example, an enterprising senator may want to gain some recognition for his abilities in energy matters. He decides to lead off the questioning in a subcommittee on oil company accounting procedures. He may begin from the premise that inventory profits represent an artificial price add-on. Thus, rather than saying, "Explain your mechanism for determining the price of the oil you have in stock," he might say, "Explain why you set the price of oil you already purchased at a pre-embargo price at the level of oil just now being delivered?" Such a line of questioning provides the journalist not only with some facts about inventory profits but with a genuine confrontation, a situation in which the questionee is put immediately on the defensive and must try to disabuse the questioner of his assumptions before proceeding with an explanation of the pricing mechanism. Although he may not fully understand the technical explanation of inventory profits, the journalist nevertheless has a story. News has, literally, been made. An atmosphere for more developments (that is, more disagreements and greater digging into "artificial pricing," as it is now known) has been established and the senator and journalist can exist in harmony for several more sessions, accomplishing their respective goals. The presumed point of the hearing—to gather evidence on what effect inventory profits have and will continue to have—is lost. So for the average reader, who still does not know what inventory profits are, there is at least the comfort that Senator X is guarding citizens' interests and forcing the uninformative, tangential oil executives to reveal corporate secrets.

THEORETICAL IMPLICATIONS

If the findings of this study are to have any practical relevance for public policy formation and journalism practice, there must be some assurances that the methodologies employed have made appropriate inquries, constituted proper techniques for seeking information, and, indeed, measured what they set out to measure. Thus, a restatement of the purposes and a review of the design of this study seem to be in order.

The study was designed in two parts. The first was a content analysis to identify the kinds of economic, political, or social analyses and sources periodicals employed in presenting monopoly or divestiture topics. The second was a questionnaire to help determine what were the self-expressed opinions of economists or journalists and whether these differing responses could be related to the review of economic literature in Chapter 1 and to the content analyses in Chapters 4, 5, and 6. The purposes of the study included identifying the concerns of both economic experts and journalists; evaluating periodical press performance in translating economic issues to their various publics; and suggesting ways that media coverage of economic issues could become more sophisticated, fairer (that is, balanced and more inclusive of relevant types of information), and more illuminating to a public or subset of publics and interest groups who, although they might possess political and social acumen, probably tend to understand little of economics in proportion to their general knowledge.

A basic hypothesis in the study is that—at this particular time and given current economic and energy supply conditions—the case against the oil industry, particularly the major companies, has been overstated and sometimes misrepresented. The complexity of examining economic issues (from the standpoint of economists, the business world, consumers, government, and so on) dictated a fairly detailed set of categories for the content analysis portion of the study. The content analysis sought a numerical counting of arguments and sources (so as to ascertain proportions of pro- and anti-industry, traditional economic, and political arguments, and proportions of types of sources); a measure of how these arguments and sources had been used (that is, in what direction, in what context, and with what intensity or valuation); a ranking of writers' attitudes or opinions (whether favorable, unfavorable, or neutral); and some verification of this ranking through the selection of illustrative excerpts from the text. However, the analysis did not rely solely on counting occurrences, for some arguments and sources did not appear.* The absent arguments or explanations were as important as those that did appear, even though they could not be counted. Nevertheless, comparisons could still be made between those publications that did or did not take certain arguments into account.

Generally, the news publications failed to take into account economic experts' interpretations of whether oil companies had violated accepted norms of competitive behavior. The business publications tended to incorporate these kinds of arguments with greater frequency (and with greater focus on economically astute, though not necessarily objective, sources), and the opinion publications exhibited varying degrees of concern with examining the economic evidence and with attaching political significance or perspective to the factual base. A limitation of the content analysis is that some of the publications' approaches are apples, others oranges. A news publication cannot make too many opinionated statements in informational articles, but in some instances, weighting of certain kinds of evidence and sources performed this function for them. The opinion publications were simply able to lend greater accent and passion to some of the same argument patterns. This is why so much attention was paid to identifying arguments, sources, direction, and frequency, for these measures enable

---

*Ole R. Holsti discusses "contingency analysis" or the notation of appearance or nonappearance of certain key symbols developed by Lasswell, Lerner, and Pool in 1952 as a supplemental measure to that of frequency. See Ole R. Holsti, "Content Analysis," in *Handbook of Social Psychology*, ed. Gardner Lindzey and Elliot Aronson (Reading, Mass.: Addison-Wesley, 1968), p. 599.

a case to be made for whether or not implicit attitudes have been reflected in seemingly impartial periodicals. Thus, the counting aspect of content analysis was thought to be a crucial aspect of making inferences about journalists' basic attitudes toward oil economics subject matter. Frequency measures of straight factual or nonattitudinal content were also felt to be necessary in order to make an analysis, which was guided by definite assumptions or hypotheses, as objective as possible. So even without the rating category of attitudes/opinions or the comments category, the study was designed to provide evidence of the manifest content of the periodicals as indications of their biases or policy preferences.

Once again, however, it should be stressed that the content analysis was largely regarded as necessary to confirm or refute this researcher's historical and economic analytical perceptions. And it was meant to be used in conjunction with the questionnaire as further indication that the periodical portion of the media had been influenced too greatly by popular arguments, had overlooked many aspects of the monopoly question, or was ignorant of some of the causes or ramifications of the issue. A positive aspect of the study is that some of the assumptions this researcher might have made about the periodical press were refuted. For example, the business publications showed a surprisingly high degree of argument and source balance, some of the news publications engaged in more complicated analysis than was originally expected, and the opinion publications revealed some concerns (for example, deceptive advertising and other consumer issues) that this researcher would have overlooked.

Another admitted limitation of this study is that it cannot measure, to any great extent, the effects periodical communications have had on general public opinion. Readers of general interest periodicals (such as *Newsweek* or the Sunday *Times Magazine*) do not come in contact with the detailed analyses of economic issues and their politically colored ramifications in the same way as do readers of opinion publications. Nor is there any practical way of separating out the messages people receive from a variety of sources—television, radio, discussions, books, other people, and so on. Thus, it would be presumptuous to suggest that periodical coverage of oil issues has caused the public to express anti-industry sentiments in public opinion polls. However, a content analysis can identify and evaluate what kinds of thoughts or arguments are being considered by diverse and far-ranging groups, institutions, or organizations. The periodicals examined here do represent a broad spectrum of views and interests. Therefore it seems feasible to extrapolate from the data as if they represented a population sample or as if a variety of societal views appeared in these publications. However, assessing the effect of periodicals' coverage of oil economics on general public opinion is outside the scope of this study.

The statistical or questionnaire portion of the study was designed to try to obtain more direct evidence of economists' and journalists' opinions on oil monopoly/divestiture. Although the content analysis alone might have provided sufficient information about journalists' perceptions of the issue, it did not tell enough about the degree or intensity of these perceptions, about the priorities journalists would assign to any divestiture efforts, or the extent to which they distinguished among monopoly problems. This last point seemed an important one because it suggests whether a journalist's assessment of monopoly is based upon consideration of economic explanations or upon political philosophy. That is, those journalists least able to distinguish among degrees of monopoly control within the stages of the oil industry tended to accept more readily the idea that the industry was anticompetitive and required external controls. One thing the questionnaire responses did illustrate is that approximately half the journalists were sufficiently unsure of the issue as to be reluctant to recommend any specific measure to curb what they believed to be the major oil companies' excessive power. The attitudes of journalists about the degree of competition, then, were more clear-cut than about the divestiture proposal per se.

There was a highly significant statistical difference between economists and journalists on the chief issue: Is there monopoly control of the oil industry by the majors? Journalists were more apt to perceive monopoly problems and less sure about which aspect of industry operations created the most severe problems. The questionnaire responses also reinforced the hypothesis that greater information fosters greater moderation of opinion. Those writers who had the greatest contact with economic or business sources tended to be more in concert with economists who had studied the issue closely. They also appeared to be more pessimistic (perhaps due to having been in the line of fire between outspoken advocates of both sides of the issue) about the media's ability to discern and present consensual economic opinion—at least on this subject.

Admittedly, the questionnaire suffered from some defects in its design. The questionnaire did not pose statements that concerned all of the respondents. It was relatively simplistic, touching lightly on the economic parameters of the issue and only superficially discussing press performance and government motives. Furthermore, the questionnaire deliberately avoided the semantic problem of whether all respondents would agree on the meaning of monopoly. (Some stressed in comments that "oligopoly" or "power" was a preferred term; one returned the questionnaire suggesting that the researcher consult a dictionary; some complained about having to rate such a "bald" statement.) However, the questionnaire was to be sent both to economists (who by training tend to view monopoly as a technical definition meaning control over price) and to journalists (who, lacking a technical definition, might regard monopoly as applying to only one firm, or as merely a loose way of saying power); thus some ambiguity was intended in order to see if the two groups would differ in their objections to evaluating the major oil companies by using this important criterion (the alleged basis for the FTC and Senate efforts). No economist registered any complaints about the use of the term monopoly, whereas journalists indicated that they really had little common understanding of this basic economic definition. A refinement in identifying economists' and journalists' views on the connotations of the term might be an improvement in studies similar to this.

Despite these limitations, the questionnaire provided strong indications of what the two groups thought about the oil industry. It also suggested that the turmoil or disagreement about the issue that has been written about in the press has exaggerated the concern the public should have about the need to alter current industry structure or modify its behavior. And it indicated that, even though both groups perceived a dichotomy on the issue among experts, the experts themselves were in concert to a large degree. The implication of this is that there has been a gap in communications as to whether some sort of consensus on the issue can be achieved. It appears that it can, but the press and the public have had a hard time seeing this.

A final theoretical issue, and one that may not be resolvable, is whether it is appropriate to expect the media to live up to a standard of expertise established by a small and, some would argue, unwordly group such as industrial organization economists. A form of further study that might suggest whether this is a fair standard would be a study of the objectivity and predictive performance or batting average of the economists identified here as experts. Although it was not thought practicable (or polite) in this study, one might wish to delve further into the biases or sympathies of academics who work closely with industrial statistics or people. Can they obtain objective, accurate information? Even so, can they be expected to interpret this material correctly, given their association with industry (especially if they have served directly as industry consultants)? Admittedly, these are questions this study has not posed. They are important sorts of questions that other researchers are encouraged to pursue.

PRACTICAL IMPLICATIONS

Implications for Public Policy Makers

This study indicates that the periodical branch of the media recorded, to varying extents, monopoly/divestiture topics or topics relating to crises in oil and gas supplies and crises in confidence in the oil industry for at least a three-year period. Yet, in this same time period, little progress was made in devising energy policies that would begin to alleviate long-term domestic supply and price difficulties. Some of the economists and journalists surveyed or mentioned in the content analysis took note of the possibility that its involvement in issuing fiery blasts against the oil industry gave Congress an opportunity to postpone making hard energy decisions. Others suggested that the oil industry had been made a convenient target (or whipping boy or scapegoat) for frustration with the OPEC oil producers or, more abstractly, the increasing difficulty with which energy could be supplied at affordable costs.

Upon examining the evidence presented by congressional leaders, federal officials, the media, and economic experts, this charge of scapegoating, although a somewhat histrionic way of putting it, seems all-in-all to express what has happened to the oil companies in the period since the OAPEC embargo. This suggests several things for public policy makers to keep in mind in evaluating past and present energy policies and in formulating new ones. One—a fact noted on several occasions even by those periodicals unsympathetic toward the oil industry—is that years of study have failed to produce evidence that the oil industry overall is noncompetitive. (To be sure, oil companies and their subsidiaries or divisions have been prosecuted for antitrust violations or for various illegal activities; but these abuses seem to have been controlled fairly well within the existing legal framework.) Thus, policy makers need to become aware of the findings and implications of past investigations and the status of current ones, but they should also be willing to accept evidence contrary to their preconceptions or expectations, once the facts are in.

A second point related to the first is that—although some information has not been ascertainable from the oil industry, and some information derived from certain companies did not jibe with data from other companies or from independent data gatherers—the majority of oil economic experts believe that the bulk of information necessary for examining the effects of industry structure and actions upon competition has been brought to the fore. Even the economists contacted for this study who claimed noncompetitiveness seemed satisfied that the data base was sufficient. Policy makers should anticipate some ambiguities, gaps, or inaccuracies in the data they receive, especially when the seek last-minute explanations in testimony or, less often, in depositions. But this should not become an excuse for making charges of secretiveness, deceit, or conspiracy.

A third point is that many a red herring has been suggested by some policy makers and swallowed whole by others. For example, Senator Henry Jackson created quite a stir by demanding to know an oil company executive's salary (in conjunction with a discussion of what Jackson regarded as "obscene" or exhorbitant oil profits). This was not an especially relevant line of inquiry. Yet the press responded to it as a newsworthy revelation. It is the job of responsible policy makers to salt heavily such behavior by colleagues and to make it clearer to the general public that they are capable of going beyond small complaints to larger issues. (In the case above, are the profits enjoyed by the major oil companies excessive beyond a temporary period during which inventory/accounting adjustments take place?) While there may be an understandable desire for policy makers to jump in and prove that progress is being made to avoid the do-nothing label, time spent making conclusions based on thorough consideration of issues in their entirety will probably be forgiven more readily than a series of dramatic but fruitless endeavors.

Fourth, since industry misbehavior had been proven in the past, and since current charges of outright collusion and the power to squeeze or create a noncompetitive market situation through production control have been difficult to prove, federal prosecutors have felt they had to redefine or discard traditional indexes of competition in order to observe the implicit effects of current oil industry operations. The idea that new measures will expose what we "feel in our bones" is a compelling one. Thus, policy makers seem to have a special obligation to be critical of the means of examining issues. They should be mindful of the objectivity (or lack of it) of certain government agencies, examine other economic evidence, and learn to hold judgments in reserve while ascertaining the factual bases of economic issues.

It now seems that many of those responsible for energy policy making and setting energy industry guidelines or regulations are coming more to grips with the interlocking nature and complexity of energy decisions. But in becoming more comprehensive in outlook, there is also the probability that they will lose touch with specific kinds of energy operations and issues (for example, environmental aspects of coal production versus electric power needs in the context of other kinds of fossil fuel production). Thus, it seems more important than ever that they be able to identify—and be willing to accept the advice of—competent specialists in particular energy fields. Policy makers, especially if they are elected officials with a multitude of responsibilities, simply do not have the expertise needed for making many detailed decisions. They are, however, trained by experience to be judges of which individuals are recognized experts in their fields, or at any rate they have staffs who could be determining this for them. In the past, at least, the Congress seems to have done an admirable job of identifying expertise. It is mainly with the use (or disuse) to which expertise has been put that this study has taken issue.

Politicians have suffered loss of esteem in the eyes of the public. Much of this could be recouped, perhaps, if greater attention were placed on the facts pertaining to issues, the studied and objective interpretation of issues, and the public ramifications of these interpretations, rather than on divisiveness, personalities, and the political implications of economic issues (for example, the them-versus-us cast given many economic issues, the pitting of hardworking senator against smug corporate bigwig, the playing up to the idea that greed forms the basis for corporate activity but never enters into the realm of public service.)

Finally, policy makers could perform a public service by calling for a reasonable attitude in examining and deciding upon energy economics issues. Although this study has made no effort to assess the influence of governmental, public-interest, industrial, and other special-interest groups, it seems likely that well-established, organized groups have been able to capture very readily the attention of the media. When they approach the media with an ax to grind, they cannot help coloring the way in which articles are written and slanted. Fervor may often be associated with truth seeking and truth seeking with investigatory prowess. Unfortunately, too many of those who have shaped government policies toward the oil industry have provided only as much evidence as was necessary to prove what they already believed. A case in point is the Senate antitrust sub-committee, the majority members of which cut off economic testimony once they had achieved the required token agreement that the major oil companies were noncompetitive. Minority members invited additional testimony from a variety of oil economic experts (some of whom were questioned in this study) and their conclusions were quite different from those of their colleagues called to testify earlier. However, few majority members stayed around to hear this later testimony.

In sum, U.S. energy policy makers, especially congressional members—if they were willing to engage in fewer displays of partisan politics, and if they were willing to share the burdens and rewards of fact finding with people who are relatively objective, knowledgeable sources in specific fields—would probably arrive much sooner at a comprehensive energy program. They also might be able to acquire greater understanding of energy economic issues if they avoided bogging down in the details others have the expertise to explore better, and if they would give the public greater understanding of the fact that economic issues are complex and require time and consideration. However, the fact that issues are complex does not necessarily mean that the public must become confused, frustrated, and hostile in order to bring about change. Change is in order only when something has gone wrong. In the case of the oil monopoly/divestiture issue, allegations were never really proved, yet members of the Senate persisted in fomenting public dissatisfaction. The public has been left with the very distinct impression that the major members of the oil industry—by their very nature being large and integrated—are sleeping giants, capable of taking advantage of the American public at any opportune moment.

Implications for Journalists

What can the press do to remedy some of the imbalance and distortion this study has revealed? Obviously, some periodicals know precisely what they are doing and intend to reinforce their reporting patterns. But for those that want to evolve and grow—and several of the publications demonstrated such an inclination—a few recommendations might be of assistance. First, although this is mere speculation, there are probably many reporters and editors who would like to enlarge their understanding of, and thus their ability to report on, economic matters. Much of the repetition of arguments and charges seen in the content analysis seems by no means deliberate, but rather, convenient. A press conference or statement is news—it gets reported. Word spreads of a news maker's expertise, involvement in an area, or ability to express him/herself in a convincing way; thus, the news maker is contacted for more quotes or explanations. He/she recommends to a reporter other people to be contacted, and often the reporter develops a beat that is philosophically unified, but generally does little to provide a broad spectrum of views. Furthermore, writers are also readers and their views, too, can be affected and reinforced by media exposure. To break out of the closed circuit of "experts," editors and writers need to enlarge or reshape continually their definition of expertise and reliable sources in any field.

The field of economics is an area in which potential sources in specialized areas are difficult for the lay person to identify, and concepts are complex and often expressed in highly technical or pedantic ways. The nuclear science field is, in some ways, more accessible (though scientists might argue misunderstood) to the average person or generalist writer than economics, in part because so many nuclear physicists have become active speakers for or against nuclear power expansion. The academic world is still the primary source of economic information today (although some academics also move in and out of government, private consulting, or corporate positions). This is not to say that noneconomists have not provided sound information in energy economics and policy ideas, nor that lay economists have not written thoughtfully on energy matters. It is to say that academic economists—at this time primarily the select group that specializes in industrial organization—are most likely to have both the talents and the inclination to study the available evidence objectively. This explains why the courts traditionally have relied on their expertise when judging antitrust cases. Furthermore, the high rate of response of this group to the questionnaire suggests that academic economists are accessible and would be willing to respond to interested media people.

How the press would actually get access to the information and/or conclusions of these experts is a more complicated matter. Certainly, greater specialization among journalism students in economics and subfields of economics would remedy some of the problems this study has identified. But even practicing journalists—if they were allotted the time and opportunity by their publications—could improve their economic awareness through formal courses or individual study (perhaps with some advice on reading materials from—catch 22—economic experts). Also, publications that have featured mainly one side of an issue could make greater efforts to find contributors with opposing views. This might require some doing, as authors might self-select away from publications whose editorial stance on an issue has been lopsided.

In addition, if editors think a study such as this has made some valid points, they could instruct writers and reporters not only to dig deeper but broader. All this, of course, hinges on time constraints, the amount of space publications can devote to special topics, and foremost, a willingness to chart more sophisticated paths in reporting. This is why a less disruptive, but more long-term solution might be to encourage writers to have greater formal training while in school. For practicing journalists, continuing education or leaves of absence are possible only if publications are willing to make greater investments in their staffs or if individuals take these efforts upon themselves.

A more uncertain issue, however, is whether—given greater comprehension of energy economic information, factors, and constraints—editors and writers might still exhibit the same biases, imbalanced argument selection, and media-genic figure focuses. That is, while some of the inadequacies in coverage this study points to may be inadvertent or nondeliberate, and others may be conscious editorial choices, still others may be the result of long-term, ingrained animosities toward corporate power or big-business practices and demeanor. For journalists, especially, some of these deep-seated prejudices may be hard to shake in order to look objectively at specific issues. The fifth estate or watchdog function of the press is a tradition that has long harbored skepticism of bigness, exclusivity, wealth. In recent years advocacy journalism, and the "new" journalism that calls for personal evaluations or involvement, have undoubtedly contributed even further to anti-industry biases (ironically, one must add, in view of the size much of mass media has attained).

Finally, there is the undeniable fact that journalists' corporate skepticism (and the desire of public officials to play up to this skepticism), coupled with persistent poking, has indeed uncovered unfair practices, greed, scandal, and illegality. In some of these cases, journalistic efforts alone revealed corporate abuses. But as the first chapter suggested, the press is often a partner (perhaps unwittingly) with government or other organized bodies in stirring anti-industry sentiment, as well as in conveying antibusiness tidings. Toeing that fine line is both a practical and philosophical problem the media have wrestled with in the past, and evidently they must continue to do so. There are no absolute standards as to how much interpreting the press should do, or how much information it should provide on the basis of its own observations. But there have, traditionally, been standards for press performance—accuracy, fairness, and balance. These rules have suffered in the reporting of the monopoly/divestiture issue; and the inflexibility of journalists' thinking is largely responsible for this failure.

If those journalists writing about the monopoly/divestiture issue had been more open to revising their views of the U.S. oil industry, a greatly changed industry since Rockefeller's day, or even since the late 1960s, there might not have been such a great disparity between public and expert thought. A major point this study has tried to illustrate is that the past does not lead inexorably to the present, and past "sins" cannot be assumed to appear reincarnated in the present. The oil industry currently illustrates such a discontinuity. Its past practices and government-granted privileges were questionable. But its current position is one over which it has greatly diminished control and which can be worsened—to the detriment of this energy-consuming society—should policy makers implement presumptive measures based upon outdated and inapplicable information. This is, in the eyes of most economists who have studied the oil industry intensively, the trap into which divestiture advocates have fallen. It is also, at least as far as this study has been able to demonstrate, the trap that snapped up so many journalists in the disturbing days since the embargo.

As the quote at the beginning of this chapter suggests, the question of who actually has expertise on any given issue will always be a difficult one to answer. However, this inquiry into the public record, press accounts, and academic studies illustrates that—in the case of the oil monopoly/divestiture issue—it has been mostly public officials and journalists, and not the experts, who have found oil industry performance not "wholesome . . . innocent . . . safe." Perhaps this study will serve as a reminder to those of us who translate complex information for general consumption that insipid common sense sometimes requires us to reevaluate how we look at, seek out, and relate what are often changing sets of facts and circumstances.

BOOKS AND CHAPTERS IN BOOKS

Adelman, M. A. *The World Petroleum Market*. Baltimore: Johns Hopkins University Press, 1972.

Allvine, Fred C. and Patterson, James M. *Competition Ltd.: The Marketing of Gasoline and Highway Robbery: An Analysis of the Gasoline Crisis*. Bloomington: University of Indiana Press, 1972 and 1974.

Berelson, Bernard. "Content Analysis." In *The Handbook of Social Psychology*. Edited by Gardner Lindzey. Reading, Mass.: Addison-Wesley, 1954.

Business History Review staff. *Oil's First Century*. Cambridge, Mass.: Harvard University Press, 1960.

DeFleur, Melvin L. *Theories of Mass Communication*. New York: David McKay, 1966.

Erickson, Edward, and Spann, Robert. "The U.S. Petroleum Industry." In *The Energy Question*. Edited by Edward Erickson and Leonard Waverman. 2 vols. Toronto: University of Toronto Press, 1974.

Festinger, Leon, and Katz, Donald, eds. *Research Methods in the Behavioral Sciences*. New York: Holt, Rinehart and Winston, 1953.

Holsti, Ole R. "Content Analysis." In *The Handbook of Social Psychology*. 2d ed. Edited by Gardner Lindzey and Elliot Aronson. 2 vols. Reading, Mass.: Addison-Wesley, 1968.

Kerlinger, Fred N., ed. *Foundations of Behavioral Research*. New York: Holt, Rinehart and Winston, 1967.

Lasswell, Harold D.; Lerner, Daniel; and de Sola Pool, Ithiel. *The Comparative Study of Symbols*. Stanford: Stanford University Press, 1952.

Mancke, Richard B. *The Failure of U.S. Energy Policy*. New York: Columbia University Press, 1974.

Mead, George H. *Mind, Self and Society from the Standpoint of a Social Behaviorist*, Chicago: University of Chicago Press, 1934.

Mitchell, Edward J., ed. *Vertical Integration in the Oil Industry*. Washington, D.C.: American Enterprise Institute for Public Policy Research, 1976.

National Research Bureau. *Working Press of the Nation.* Edited by Milton Paule. 4 vols. Burlington, Iowa: National Research Bureau, 1976.

Osgood, C.; Suci, G.; and Tannenbaum, P. *The Measurement of Meaning.* Urbana: University of Illinois Press, 1957.

de Sola Pool, Ithiel, ed. *Trends in Content Analysis.* Urbana: University of Illinois Press, 1959.

Stone, Phillip J.; Dunphy, Dexter C.; Smith, Marshall S.; and Ogilvie, Daniel M. *The General Inquirer: A Computer Approach to Content Analysis.* Cambridge, Mass.: M.I.T. Press, 1966.

Tarbell, Ida. *The History of the Standard Oil Company: Briefer Version.* Edited by David M. Chalmers. New York: Harper and Row, 1966.

Twentieth Century Fund Task Force on United States Energy Policy. *Providing for Energy.* Background paper by Richard B. Mancke. New York: McGraw-Hill, 1977.

GOVERNMENT DOCUMENTS

Federal Trade Commission. *Complaint in the Matter of Exxon Corporation; Texaco, Inc.; Gulf Oil Corporation; Mobil Oil Corporation; Standard Oil Company of California; Standard Oil Company (Indiana); Shell Oil Corporation; and Atlantic Richfield Company,* Docket No. 8934, 18 July 1973.

U.S. Cabinet Task Force on Oil Import Controls. *The Oil Import Question.* Washington, D.C.: Government Printing Office, 1970.

U.S. Congress, Senate, Committee on the Judiciary, Subcommittee on Antitrust and Monopoly. *Government Intervention in the Market Mechanism: Hearings before the Subcommittee on Antitrust and Monopoly of the Committee on the Judiciary on S. Res. 40,* 91st Cong., 1st sess., 1969.

U.S. Congress, Joint Economic Committee, *Hearings before the Joint Economic Committee,* 94th Cong., 2nd sess., 1975.

U.S. Congress, Senate, Committee on the Judiciary, Subcommittee on Antitrust and Monopoly. *Economists' Testimony: Hearings before the Subcommittee on Antitrust and Monopoly of the Committee on the Juciciary,* 95th Cong., 1st sess., 1976.

JOURNAL ARTICLES

Anderson, C. David. Review of *Competition Ltd.: The Marketing of Gasoline,* by Fred C. Allvine and James M. Patterson. *Yale Law Journal* 82 (May 1973): 1355-62.

Erickson, Edward, and Spann, Robert. "Entry, Risk Sharing and Competition in Offshore Petroleum Exploration." North Carolina State University, December 1975. Xerox copy.

McGee, John S. "Predatory Price Cutting: The Standard (N.J.) Case." *Journal of Law and Economics* 74 (Fall 1958): 137-69.

McKie, James. "Market Structure and Uncertainty in Oil and Gas Exploration." *Quarterly Journal of Economics* 74 (November 1960): 543-71.

Stigler, George. "Monopoly and Oligopoly by Merger." *American Economic Review* 50 (May 1960): 30-35.

ABOUT THE AUTHOR

BARBARA HOBBIE is associated with the Program on International Energy, Resources and the Environment at the Fletcher School of Law and Diplomacy, Tufts University.

Ms. Hobbie has also served as an editorial consultant to a number of non-profit organizations and foundations and previously coordinated the publications program for the Educational Resources Information Center/Counseling and Personnel Services, a National Institutes of Education project.

Ms. Hobbie holds a B.A. in English and an M.A. in Journalism (Communications Research Sequence) from the University of Missouri at Columbia.

RELATED TITLES
Published by
Praeger Special Studies

ENVIRONMENTAL LEGISLATION: A Sourcebook
edited by Mary Robinson Sive

*THE MEDIA AND THE LAW
edited by Howard Simons
Joseph A. Califano, Jr.

PERSPECTIVES ON U.S. ENERGY POLICY:
A Critique of Regulation
edited by Edward J. Mitchell

*THE UNITED STATES AND INTERNATIONAL OIL:
A Report for the Federal Energy Administration on
U.S. Firms and Government Policy
Robert B. Krueger

*Also available in paperback as a PSS Student Edition